THE POWER OF PLANKTON

The Power of Plankton

How Plankton Made Life on Earth Possible and Why It's Key to Our Future

Vincent Doumeizel

Translated by
Charlotte Coombe

THE **BOOK SOCIAL**

THE BOOK SOCIAL, AN IMPRINT OF LEGEND TIMES GROUP LTD
3rd Floor
86-90 Paul Street
London EC2A 4NE
www.thebooksocial.co.uk

First published in French as *Le Manifeste du plancton* by Éditions des Équateurs/Humensis in 2025
This translation first published by Legend Times in 2026

This book is supported by the Institut français (Royaume-Uni) as part of the Burgess programme.

Printed by Akcent Media, 5 The Quay, St Ives, Cambs, PE27 5AR

ISBN: 978-1-91829-198-8

To M, N and E.

'What is essential is invisible to the eye.'
Antoine de Saint-Exupéry,
The Little Prince

The Power of Plankton

Introduction

To understand plankton, let's start by observing a droplet of water. Because it's transparent, we think it's empty. Actually, it is an abundant ecosystem teeming with life. Millions of invisible organisms live together and interact inside it. This is where an extraordinary story began: the unique epic that is life on earth.

Around 3 to 4 billion years ago, tiny organisms began drifting around in the ocean. Much later, they would come to be called *planktos* – meaning 'drifting' or 'wandering' in Greek – to describe their erratic behaviour. They are characterized by the way they live in suspension and are unable to swim against the current that carries them. These drifters also happen to be our ancestors. We, just like our goldfish, are descended from plankton.

Fairly recently in the history of life, plankton organisms began to organize themselves, sharing tasks and specialising so as to increase efficiency and help each other. This is what we now call multicellularity, which was a major turning point in the history of life. Then, these multicellular organisms left the ocean to conquer the continents and develop terrestrial life. Plankton is therefore the source all life on earth and the world as we know it. The root of the word 'plankton' is the same as that of the word 'planet'. Plankton drift randomly in the ocean, just as our planet does in the universe. Our earth is historically and biologically a 'plankton planet'.

Of the 4 billion years of evolution of living organisms, 3.5 billion took place in the ocean alone, and for the most part in planktonic form. The history of life is therefore 90 per cent marine, and terrestrial life is only a recent epiphenomenon. Furthermore, although the ocean covers 71 per cent of our planet's surface, it accounts for 96 per cent of its habitable space. In fact, a large number of living creatures live in the ocean at an average depth of 3,700 metres. For the most part, these are plankton.

Some of them belong to the category of plants, others to that of animals, and many to both, when they are not viruses, fungi or bacteria.[1] Different types of plankton are responsible for the colour of salmon scales, flamingo feathers and prawn shells, as well as the ocean, the Red Sea and blood rain.[2]

Amazingly, it is one of the least understood ecosystems, but it is also our planet's most important: it accounts for 95 per cent of the biomass in the ocean. But what exactly is this enigmatic creature, whose blooms of life are visible from space?

Plankton includes organisms as varied as bacteria, micro-algae, tiny shrimps, fish larvae and even jellyfish. Its biodiversity is so vast that it includes viruses 5,000 times finer than a hair and siphonophores 100 metres in length.[3] In terms of scale of size, if the smallest plankton was the size of an ant, then the longest would be the length of Great Britain![4]

Fish, invertebrates,
mammals, marine vegetation
5 %

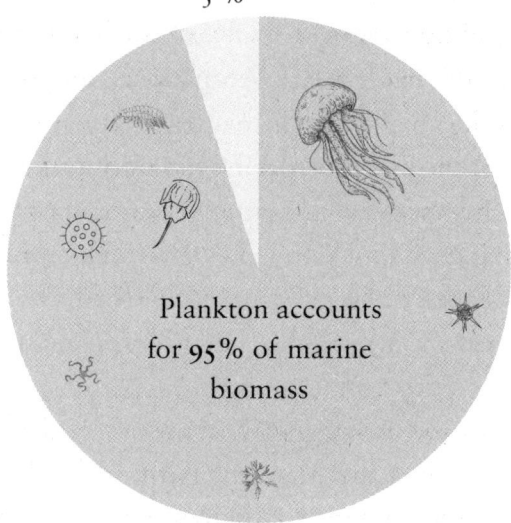

Plankton accounts
for 95% of marine
biomass

Distribution of marine biomass
Source: Sciences Participatives, MNHN

How did these nomadic organisms create such diversity and, above all, help life to evolve on our planet? Even before they had limbs that facilitated movement, their ability to drift, and therefore to get around, enabled them to meet each other. Some of these beings devoured each other, many helped each other, others merged, but they all innovated so as to adapt to external conditions. In this way, they accelerated evolution.

Their first stroke of genius was to create photosynthesis, one of the most important inventions in the history of life.[5] Plants simply imitated their plankton ancestor by using the sun's energy to transform carbon dioxide into living matter while releasing oxygen.

As plankton drifted and moved around, feeding on the light and nutrients floating in the ocean, they also transformed their environment and invented the geochemical cycles of the main elements. And so, our mysterious protagonist is the source not only of the oxygen we breathe, but also of the carbon, nitrogen, phosphate and freshwater cycles, and even of cloud formation, rain and the entire water cycle on planet earth.[6] When they died, these microscopic organisms settled on the bottom of the ocean. Over millions of years, their shells agglomerated to form huge geological layers up to several kilometres thick.[7] When major tectonic shifts pushed the continents out of the ocean, these billions of dead plankton shaped many of the rocks we know today. The cliffs at Dover, the pyramids of Giza and Europe's ancient cathedrals are all built from plankton.

Finally, since its origins, some of the dead plankton slowly decomposed and penetrated the soil, producing the oil and gas we now consume in such large quantities.[8] These essential characteristics make plankton a clear ally in tackling the ecological threats that lie ahead for us...

At a time when we are worried about the fate of turtles, dolphins, large fish and whales, we urgently need to take a closer look at what feeds them: plankton. It is at the base of the aquatic food chain and enables all marine life to exist. According to the Food and Agriculture Organization, 3.3 billion people depend on fishing and other marine products.[9] To avoid global famines, we need to be aware that the food security of our societies depends largely on the abundance of plankton.

Not only is it the source of all life, but it also plays a vital role in the sustainability of our planet. Plankton can contribute to slowing down climate change and sequestering some of our carbon emissions by falling as 'marine snow' to the bottom of the ocean.

The genetic potential for innovation is still virtually unexplored, but plankton may soon be able to feed our populations, clean our rivers, regenerate biodiversity, ingest the ocean's plastic, treat our sick and provide us with light. In *Twenty Thousand Leagues Under the Sea*, Jules Verne describes how myriads of luminous animalculae glide over the hull of the *Nautilus*.[10]

My interest in seaweed led me to meet many seaweed farmers who have been invaded by invisible enemies, planktonic blooms, whose proliferation could wipe out their entire production in just a few days. However, a healthy balance of plankton can also prove to be beneficial for our aquatic plants. In turn, the restoration of kelp forests promotes an abundance of plankton and of the entire ecosystem that depends on these organisms.

For all these reasons, there is an urgent need to better understand this essential 'black box' that supports – and can sometimes destroy – life underwater and on land. Under the aegis of the United Nations, we brought together around thirty plankton specialists from all over the world to work together on a short document setting out the role, importance and actions to be taken on the subject. For the first time, oceanographers, ecologists, marine biologists and historians have drawn up a collective report designed to raise public awareness: 'The Plankton Manifesto'.[11] This twenty-page manifesto was published in September 2024 at the UN General Assembly in New York. It has since been used as a reference tool at major institutional meetings. This book is a continuation of that manifesto.

The world of plankton is really only just beginning to become visible to us. Thanks to new tools such as DNA sequencing, satellites and high-definition electronic microscopes, it is now possible to understand this immense and essentially microscopic biodiversity. Until now, it has remained distant and invisible. It is a bit like nobody on earth knowing anything about forests. A forgotten civilization, long predating ours, emerges from the shadows, like the discovery of a sunken Atlantis.

Understanding plankton is now an absolute priority if we want to preserve biodiversity and combat global warming. It has the potential to reverse the fate of the planet, as it has done several times in the past. If we preserve it, plankton will save us. On the other hand, if we continue to disregard it and damage the balance of ecosystems, many forms of life, including our own, will disappear.

Alain Bombard: overcoming ship-wrecks with plankton and hope

In 1951, the young French doctor Bombard was unable to save the victims of a shipwrecked fishing trawler in Boulogne-Sur-Mer. Devastated by this event, he decided to dedicate himself to ensuring survival at sea. At the time, 200,000 people died every year from shipwrecks; 50,000 of them in lifeboats. The doctor identified three possible physiological causes of death: cold, thirst and hunger. Death from cold will occur after a few minutes, death from thirst after three days and from hunger after three weeks. However, most deaths occur within one to two days, so are therefore not linked to any of these causes. His research into past shipwrecks, particularly the *Titanic*, shows that survival rates are higher for children than for adults, despite having lower metabolic reserves. In reality, the cause of these deaths is not physiological but psychological. It is despair and fear that kill. Children are less aware of the dangers. They tend to remain hopeful.

So the remedy is simple: raising the level of hope among those who are shipwrecked. Bombard then decided to do an experiment to demonstrate that survival was possible when spending dozens of days in a dinghy lost at sea. Obviously, he couldn't find anyone to volunteer for such an experiment, so he decided to do it himself. After a trial in the Mediterranean, he embarked on a solo Atlantic crossing in a rubber dinghy that he christened *L'Hérétique* ('The Heretic'). On 19 October 1952, gazing out to the horizon, Bombard left the Canaries and headed for the Americas.

He managed to defeat the cold by putting on layers of clothing. To avoid dying of thirst, he collected rainwater, his own urine and seawater. In order to feed himself, he had a plankton net. The ocean he was sailing on was full of planktonic organisms. So, during this long drift across the Atlantic, he filtered the seawater every day to recover these tiny crustaceans, known as copepods. Zooplankton is rich in lipids, vitamins and fatty acids, so he only needed a small amount to meet his daily needs and prevent scurvy. Although this 'slimy, gooey sludge' was not particularly appetising, it did provide him with the minimum nutritional requirements.

The days went by – ten, fifteen then twenty. During that time, all he had for company was a little cuddly toy. The ordeal was becoming more and more difficult with storms, foot injuries, raw flesh eaten away by salt, temperature variations, falling blood pressure, diarrhoea, weight loss, lack of rain, risk of drowning and fear of death... The challenge was turning into a nightmare. After just over forty days, exhausted and desperate, he wrote his will, admitting failure. But then he pulled himself together. Using his plankton net, he started filtering water again. Luckily, a cargo ship picked him up and offered him a meal. He then realized that his inaccurate calculations in handling the sextant had caused him to drift 600 miles (1,000 km) from the optimum course he had planned. He thought he was almost at his destination, but in fact he was still in the middle of the ocean with several weeks' worth of sailing still to go. Despite the state he was in, Bombard refused to give up. Like a man possessed, under the incredulous eyes of the captain, he climbed back on his inflatable, alone and exhausted. The only weapons he had left were a plankton net and his optimism.

On 23 December, after 65 days at sea, he landed in Barbados. He had lost 25 kg since setting off. His book *Naufragé volontaire*, originally published in 1953 and later published in English as *The Bombard Story* or *History of a Voluntary Castaway*, was a global success. Bombard even became a generic name for a lifeboat. Until his death, the indomitable doctor-navigator received thousands of letters of thanks from shipwreck survivors. Without his example, they are certain they would all be dead. Like Bombard, we need to cultivate optimism in the face of adversity, because fear and panic kill us more than the actual dangers. Hope saves us. The global state of the planet can be compared to this formidable journey. Rather than get discouraged by the tragedy threatening the whole living world, we should try to gain a better understanding and find new solutions.

Plankton and hope are the two common threads of this book. This fresh hope comes to us at a time when the world seems to be heading for a major ecological and climatic disaster. Although invisible, plankton is essential to life on earth and has played a crucial role in its equilibrium for billions of years.

Arise, plankton citizens!

Plankton is notably absent from national and international political discussions, even though it supports the majority of the world's biodiversity. And it is mentioned even less when it comes to climate discussions. The ocean absorbs 30 per cent of our emissions, mainly thanks to plankton. Yet its major role is still widely ignored. The first mention of the ocean in a COP Climate agreement dates back to the Paris Agreement at COP21. So, 21 COPs had taken place before we started showing any interest in the thing that absorbs a third of our emissions. By the time the Climate Conference in Brazil took place in 2025, still no mention had been made of the main contributor to this carbon pump: plankton.

Our total disregard of plankton starts at an early age. Despite the importance of plankton in our history and in the balance of our planet, we never study it at school. It is totally absent from our collective imagination. At best, we associate it with krill, or with mysterious invisible organisms floating in seawater, or even with the dastardly microcrustacean in a well-known cartoon whose hero is a sponge. We are often unaware of its role in the earth's delicate balance, even though we have a treasure floating in abundance in the oceans right before our eyes.

How many of us could actually draw a plankton? Is it

an animal, seaweed, plant or crustacean? And what is its biological role?

Yet it would be fun and exciting to grow plankton in the classroom and carry out experiments to see, in real time, the incredible speed at which these organisms reproduce and how they interact with their environment. We could admire plankton radiating light in the water, study the splendid plates by the nineteenth-century German biologist Ernest Haeckel, Christian Sardet's incredible photographs against a black background, Lucile Viaud's sculptures in sea glass made from plankton, Ryoji Ikeda's plankton-inspired multimedia forms, or sample Angel León's delicious plankton sauce.

In April 1970, the French children's magazine *Pif Gadget* distributed packets containing 'pifises' which were the dehydrated eggs of *Artemia salina*, a plankton that, as its name suggests, is particularly fond of saltwater. The idea was for people to hatch these eggs and watch how they transformed into microcrustaceans. One million copies of that issue were printed, a record in the history of French comic magazines. Building on this success, the planktonic gimmick was reused in issues from 1973, 1977 and 1981. Which just goes to show that plankton *can* have mass appeal.

Unfortunately, interest was short-lived and remained nothing more than a bit of childish fun for the magazine's young readers. Even in higher education, plankton is still an enigma, somewhere between molecular biology, oceanography and marine biology. The disconnect between the ecosystem importance of plankton and the lack of awareness of it in our culture starts at a very young age. There are too few adults – and particularly researchers – specializing in plankton. It's still a largely unexplored field, but one that holds almost limitless resources for our future. In France today, a number of pioneers such as Tara and Plankton Planet, supported by sailors and scientists, are launching initiatives to study plankton and raise awareness of it. In the same vein, the Océanopolis museum in Brest has launched a major participatory research initiative dedicated to the study of coastal plankton, called 'Objectif Plancton'. Since 2024, even the French Navy – through the Bougainville Mission – has agreed to use its military vessels to collect samples of different types of plankton around the world. Many other initiatives could be mentioned, notably in UK, USA, China and Japan. Understanding plankton is everyone's responsibility. The sampling of our aquatic spaces depends on each and every one of us. We all have to do our bit as plankton activists.

In Port-Louis, the biologist Pierre Mollo, a great storyteller of plankton, launched a 'plankton observatory' in 2003 offering training courses, events and workshops about these microorganisms. A wonderful virtual reality initiative, Plankto-Quest, was recently launched by a group of marine biologists from the CNRS and the University of Grenoble. A virtual reality headset enables you to immerse yourself and interact in 3D with underwater plankton, handle them and even feed them. Another innovation developed by researchers at the CNRS in Villefranche-sur-Mer is the *Play For Plankton* game, a plankton version of the popular *Candy Crush*. Set in an aquarium, the aim is to sort plankton on the basis of their similarities in order to classify them more quickly in a 'planktopedia'. Once they've all been sorted, you move on to the next level. The app, which is available on all smartphones, has been a great success in helping to fund research, as all the photos come from samples taken by scientists. By playing the game, users are helping researchers to classify millions of images of microscopic organisms.

In 2011, UNESCO supported the series *Plankton Invasion*, broadcast on Canal+, while Netflix released an animated film, *Plankton: The Movie*, in 2025, based on the baddie in *SpongeBob SquarePants*. Whether it's a question of purely scientific research or education, funding, or introducing new types of narrative that encompass all living things, the time has come to tell the story of one of our planet's most incredible epics. Each and every one of us needs to become a responsible citizen of our 'plankton planet'.

LUCA, our very first ancestor

In the nineteenth century, Darwin outlined his theory on the evolution of species. However, the true root of our genealogical tree remains shrouded in mystery. It all began 4.5 billion years ago, when the earth was forming in a chaos of stardust and magma, and collided with the planet Theia. The moon was born out of this collision. Then, a cascade of meteorites ripped through the surface of our planet. It was probably around this time that water first appeared on earth. A necessary condition for life, we still know almost nothing about how it appeared. Was it a shower of icy meteorites, or global cooling that turned gases into liquids?

In the middle of the Hadean era – the first geological era in the history of the earth, named after Hades, the god of the underworld in Greek mythology[12] – the earth was a living hell. Its crust was riddled with submarine volcanoes spewing gases of primitive compounds: hydrogen, methane, nitrogen, ammonia and sulphur gushed out at insanely high temperatures. Despite this hostile climate, between lava flows and the formation of the first oceans, a living cell capable of reproducing appeared. This cell, named LUCA *(Last Universal Common Ancestor)* is thought to be the hypothetical common ancestor of all forms of life, including planktonic microorganisms (bacteria, archaea, eukaryotes). So LUCA is the ancestor of us all!

Its appearance is still a mystery. Recently, some scientists have speculated that it may have come from space, transported frozen inside meteorites. The standard hypothesis, however, is that it came about near hydrothermal springs in the depths of the ocean. These extreme environments – heat, pressure, absence of light – would have encouraged complex chemical reactions leading to the formation of primitive cellular structures.

Recent genetic studies estimate that it appeared around 4.2 billion years ago, not long after the earth was formed.[13] But more importantly, LUCA was not alone. It existed surrounded by other organisms on the boundaries of life, viruses and other cells that have not, to this day, found a path to life. LUCA developed in opposition to them and thanks to them. Who are these ghosts that lived alongside our ancestor? We'll never know. LUCA was not the first autonomous organism. It was one of many inhabitants of its microscopic world, but significantly, the only one to have descendants that reached us.

LUCA's descendants managed to leave the marine sediments and become plankton. They learned to live and, above all, to reproduce, suspended in water (which is no small feat). In their wandering and their drifting, they conquered the world! This mobility allowed them to discover other organisms, other living conditions, to mutate, to adapt and to evolve.

And so, within a few billion years, LUCA and its nomadic marine descendants achieved this unique feat: transforming stardust into a biosphere.

Plankton that came from space and might soon return?

Some of the water in our oceans is older than our planet and our solar system.[14] Where does it come from? I asked myself this question while visiting a European Space Agency (ESA) site where they are working on sending plankton into space. The work was focused on a well-known type of planktonic bacteria: spirulina.

Far from being limited to organic stores and food supplement enthusiasts, this miraculous freshwater plankton is generating a great deal of interest in the space sector. This cyanobacterium turns out to be an ideal food that is extremely rich in nutrients, easily transportable, perfectly suited to long-term space missions. It also plays an essential role in the absorption of carbon dioxide and the production of oxygen, a crucial parameter in the context of space travel where it's not really advisable to open the porthole for ventilation. Space agencies around the world, in Europe, the United States and Russia, have been interested in it for decades.

Another advantage: spirulina grows only under the effect of light, which allows production to be increased or decreased as required. The process of photosynthesis cannot happen without nitrogen, but once again, the solution is onboard. Urea produced by humans is rich in ammonia, and therefore in nitrogen. Urinating on the bacteria promotes their productivity. The European Space Agency has thus set up experimental sites in Barcelona and the Netherlands. It even sent tubes filled with spirulina to the ISS (International Space Station) to validate the growth levels of microalgae in such a particular environment (no gravity, cosmic radiation, etc.).

The project is in fact more wide-reaching. These extraordinary bacteria offer a very high reproduction rate, resistance to radiation and an exceptional history of adaptation to extreme conditions. This makes them excellent candidates for creating oases of life for populating new planets. After all, they have already filled our atmosphere with oxygen. No doubt they can repeat that elsewhere. To conquer space, it therefore seems wiser to take these types of bacteria than to take trees.

Although I found spirulina fascinating, I still didn't have an answer to my initial question. Where does our planet's water come from? And more importantly, could it have transported the bacterial life forms that later established life on earth? This possibility would give ESA studies incredible depth. Imagine billions of frozen, water-filled projectiles crossing the infinite void of the universe and crashing into our planet at the dawn of its existence. This icy water may have helped create the ocean. If this hypothesis were verified, it could explain LUCA. An extraterrestrial virus carried in the frozen water of an asteroid that is believed to have collided with earth. Some people believe this. The first traces of life 4.2 billion years ago, only 300 million years after the creation of our planet, make this hypothesis probable. A few hundred million years in such a hostile environment seems like a very short time to initiate a project as complex as life.

The case is not unique. A few hundred million years later, the living world experienced the sudden emergence of a particular type of bacteria capable of simultaneous photosynthesis and nitrogen fixation. We will come back to these cyanobacteria, which produced oxygen on earth, later. Could they have also come from space?[15] Some renowned researchers think so. These mutations, which have no intermediate evolutionary trace, might be of extraterrestrial origin. The first traces of photosynthesis on earth date back 3.5 billion years to a time when aquatic life on Mars may have existed, as the latest NASA exploration missions seem to confirm.[16] Could it be that bacteria survived a cataclysm on Mars to reach earth? According to this hypothesis, we are all descended from Martians.

If we take a step back and look at the history of bacteria as a whole, we see a genuine form of bacterial destiny. Because if our bacteria came from space 4 billion years ago, perhaps they will soon end up returning there thanks to a creature they created: the human being. More and more researchers consider humans to be essentially bacterial beings. Our bodies, with their microbiota, contain ten times more bacterial cells than human cells. Bacteria shape our lives and probably many of our decisions.

So bacteria may have crossed galaxies to land on earth. In any case, they filled our atmosphere with oxygen and colonized the earth through mutations that happened as they drifted in the ocean. Then, becoming microbiota in our organs, they decided to use human exoskeletons capable of manufacturing space vessels to resume their route towards the Milky Way. And to go and settle on other planets... Who needs science fiction, when we have life?

Ernst Haeckel: the father of ecology, and a source of inspiration for the Nazis

In 1900, at the Paris Exposition, the visitors all noticed the spectacular entrance door designed by the great architect René Binet. It was inspired by a plankton drawn by a renowned German biologist who is very controversial today: Ernst Haeckel.

A plankton enthusiast, famous scientist and great friend of Darwin, still celebrated by some today as the 'father of ecology', Haeckel is an ambivalent figure. Even though he was a Freemason who was close to the social democrats and a fervent pacifist aware of issues relating to feminism and homophobia, at the very beginning of the twentieth century he supported theses recognizing the supposed superiority of the 'white race' and was an honorary member of the German 'Society for Racial Hygiene'. He also notoriously falsified drawings to support certain scientific theories that were later proven to be false.[17] Some have called this episode 'the most famous falsification in the history of biology'.[18]

Born in 1834 in Potsdam, the young man was torn for a long time between a career as an artist and one as a doctor, before opting for the latter path. Leaving medical practice to pursue biology, he started combining life sciences with art, the two disciplines being, according to him, very closely linked. He thus developed his theory of 'the great whole' at the origins of ecology, a century before environmentalism became a political movement. Haeckel also referred to politics as 'applied biology'. In Greek, ecology means 'the study of habitat': it is about understanding the interactions between living organisms and their environment.

In 1866, aged thirty-two, Ernst Haeckel visited Charles Darwin, who was twenty-five years his senior. Seven years earlier, the naturalist had published a work that revolutionised the science of life, *On the Origin of Species*. The two scientists exchanged ideas for twenty years with great mutual respect. Haeckel, deeply influenced by these exchanges, developed the idea of a common origin of life. He is therefore the inventor of LUCA. It was in the context of his reflections that he began to draw his first 'trees of life', which were to have a huge impact on the history of biology.[N]

Highly acclaimed by leading figures, including the writer Flaubert, who lauded his talents for popularization, he was also one of the first to envision the birthplace of humanity in Africa, after earlier hypothesizing its origins as being in Southeast Asia or a submerged continent. Haeckel was particularly fascinated by the shapes and nature of plankton. His incredible drawings help make these microscopic species known to the general public. Hundreds of plates from his book *Art Forms in Nature* are still widely referenced in biology to this day, both for their level of detail and for the diversity of the species represented, not to mention their aesthetic appeal. According to Haeckel, photography cannot equal drawing when it comes to the precision and complexity of forms, and the vividness of colours.

Haeckel travelled the world, studying and drawing thousands of marine species and plankton, from radiolarians to jellyfish and siphonophores. He was so fascinated by jellyfish that he decorated his house with engravings of these tentacled, gelatinous beauties and even named it 'Villa Medusa'.

His scientific contribution, particularly to our understanding of plankton, was immense, but remains horribly tainted by his ideas on 'social Darwinism', an ideology that uses the idea of natural selection to justify social, economic or racial inequalities in human societies. Many researchers who still reference him argue that his ideas must be viewed within the scientific context of the time, that the huge complexity of his ideas cannot be reduced to that alone, and that, while some of his concepts were taken up posthumously by the Nazis, others were radically rejected.

He nevertheless remains a divisive figure because of his unacceptable racist theories. Haeckel's contribution to research was enormous, but he also made significant mistakes and embraced ideas that were the opposite of those expected of such a scholar. The same as for artists and writers, scientific genius does not excuse everything.

Trans-animal and trans-vegetal plankton

The current concept of the 'fluidity' of beings could have its origin in the ocean. There, the notion of 'trans' does not only apply to gender but calls into question the distinction between the great kingdoms of living things. Traditionally, life is divided into autotrophic plants (which use photosynthesis to produce their organic matter), heterotrophic animals (which feed on plants or other animals to produce their organic matter), and bacteria and fungi (which break down matter to recycle it); all regulated by viruses. But in the ocean, this classification is shattered.

All these groups are found there, but in very different assemblages. First, life is divided between those that live on, in or near the seabed (benthic zones) and those that inhabit the open water column (pelagic zones). Among the latter, there are nekton that can actively swim against the current, and plankton that drift. Plankton includes plants and marine animals, but above all an immense majority of hybrid beings, neither totally animal nor totally vegetal, but a bit of both at the same time: the 'trans-kingdom'.

For a long time, science classified these organisms by analogy with terrestrial models: photosynthetic organisms were assimilated to plants and called 'phytoplankton', while those that consumed organic matter were called 'zooplankton'. But in the 1990s this classification was called into question (although it is still used). Many organisms are in fact capable of both photosynthesis and absorption from other cells. These strange organisms belong to a hybrid category possessing both animal and plant characteristics, which could be called 'vegimals' and are sometimes referred to as 'animal+vegetal'.[19] The scientific term, however, is 'mixoplankton'.

In this book, we will use the terms 'primary producers' or 'photosynthetic plankton' to include cyanobacteria, phytoplankton and mixoplankton. Often popularized under the term 'microalgae', mixoplankton constitute the majority of plankton, and can reach concentrations ranging from 100,000 to 100 million individuals per litre of water!

The term plankton actually represents an extremely diverse group, ranging, for example, from a 100-metre-long siphonophore to a planktonic virus of a few nanometres in size. This heterogeneity is also found in zooplankton, which includes permanent planktonic species and the larvae of marine animals destined to become fish, crustaceans, shellfish or molluscs.

In the oceans, the basis of the food chain is photosynthetic plankton, which can transform CO_2 into organic matter. Its production rate is impressive: 50 per cent of the world's photosynthetic plankton is renewed every week. Every day, plankton produces seven to ten times more organic matter than the Amazon rainforest![20]

The marine trophic pyramid illustrates this dependence: 1 tonne of plankton feeds 100 kg of zooplankton, which feeds 10 kg of crustaceans or fish, which feed 1 kg of anchovies or small fish. These can feed 100 g of large predators such as tuna, salmon or shark. So when you consume 100 g of tuna, you indirectly ingest a tonne of phytoplankton!

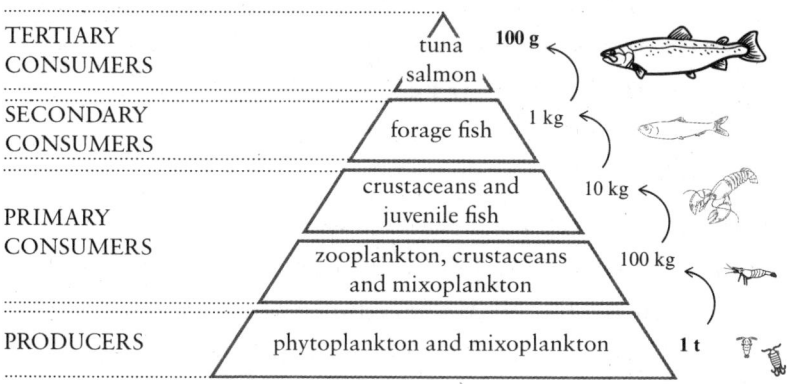

Simplified pyramid for the production of 100 g of large predators from photosynthetic plankton[*]

If a single link in this chain is missing, the marine balance collapses like a house of cards. Understanding plankton dynamics is essential for fishermen, but also for oceanologists and ecologists. The abundance of fish and marine life depends entirely on these organisms that are invisible to the naked eye but absolutely fundamental to ocean life.

---

[*] This pyramid represents the trophic chain over a long period, because at a given time, we would have an inverted pyramid. More on this later.

The first mass extinction: a toxic waste called oxygen

Around 3 billion years ago, strange blue bacteria called 'cyanobacteria' brought about a radical shift in the trajectory of the living world. These direct ancestors of spirulina used the sun to create living matter. This is called photosynthesis, arguably the most revolutionary innovation in the history of life on earth. Microscopic cells, surrounded by a membrane 50,000 times thinner than a hair, were solely responsible for this amazing process that uses carbon and nitrogen to transform sunlight into living matter. All without a nucleus, without a single neuron, and just with a single cell and a single chromosome! No one has managed to top this since then. With the sun as their sole source of energy, these new types of cells did everything they could to get closer to the surface. Floating in the ocean, they became plankton.

The process of photosynthesis allowed them to break down carbon dioxide (CO_2) dissolved in water to recover the carbon. But their brilliant invention caused collateral damage, producing an unexpected and incredibly destructive 'waste product' for existing life: oxygen.

At the time, the earth was populated only by 'anaerobic' organisms; the creation of living matter happened thanks to methane and sulphur. In that world, oxygen was a highly toxic gas. For several million years, this oxygen was fortunately captured by the liquid iron present in the ocean. On contact with this gas, the iron oxidised, captured oxygen, solidified, then settled on the seabed. This chemical reaction transformed the planet into a scarlet sphere. A land of rust! This episode, the Great Oxidation Event (GOE) is at the origin of the immense iron deposits that remain a source of wealth for South Africa, Brazil and Australia.

Unfortunately, the method invented by our cyanobacteria was so effective that they proliferated. So much so, that their oxygen ended up consuming all the available iron. Freed from its cage, oxygen began to spread freely into the ocean, then into the atmosphere. This climatic upheaval systematically eradicated nearly two billion years of evolution. Practically no organisms escaped this cataclysm, not even our cyanobacteria. Only a few plankton managed to take refuge in marine sediments or in the deep sea, the only places still protected from this horrific gas.

By destroying life with their oxygen, cyanobacteria profoundly modified the climate because the anaerobic organisms populating the earth at the time produced methane, the main greenhouse gas maintaining the earth at high temperatures. When they disappeared, the planet suffered a sudden drop in temperature (around -25°C), leading to a global cooling: an ice age known as the 'Huronian Glaciation'. For nearly 300 million years, the earth lost its rusty iron colour, and put on an immaculate white coat of ice.

Despite this second catastrophe, some plankton managed to survive and adapt. Slowly, new equilibriums formed. Certain bacteria started producing methane again, gradually warming the planet. Other more evolved bacteria learned to use oxygen for their metabolism, creating cellular respiration. This process marked a new turning point: more complex organisms appeared, such as algae, plants, and finally, animals. They all share a common heritage, as they are the result of an adaptation to a major ecological disaster.

Oxygen, which was once a symbol of death for primitive organisms, became the foundation of life on earth. This ancient event offers a striking parallel with our times. For several decades, humans, as aerobic organisms, have been emitting massive amounts of carbon dioxide and methane, and disrupting the earth's climate. These gases released by human activities reverse the natural process that once cooled the planet, so it is now warming up. This anthropogenic phenomenon threatens the balance of life just as the Great Oxidation Event did over two billion years ago. The history of cyanobacteria shows that such climatic upheavals can wipe out entire ecosystems but can also open the way for new forms of life.

Today, the question remains: will humanity be able to curb this destructive dynamic and prove itself to be smarter than primitive bacteria? The essential difference is that we're now fully aware and informed of the situation. A better understanding of the role of plankton – key to climate regulation – could offer solutions to limit the damage. By integrating the lessons of the past, it is still possible to prevent ourselves from reliving it.

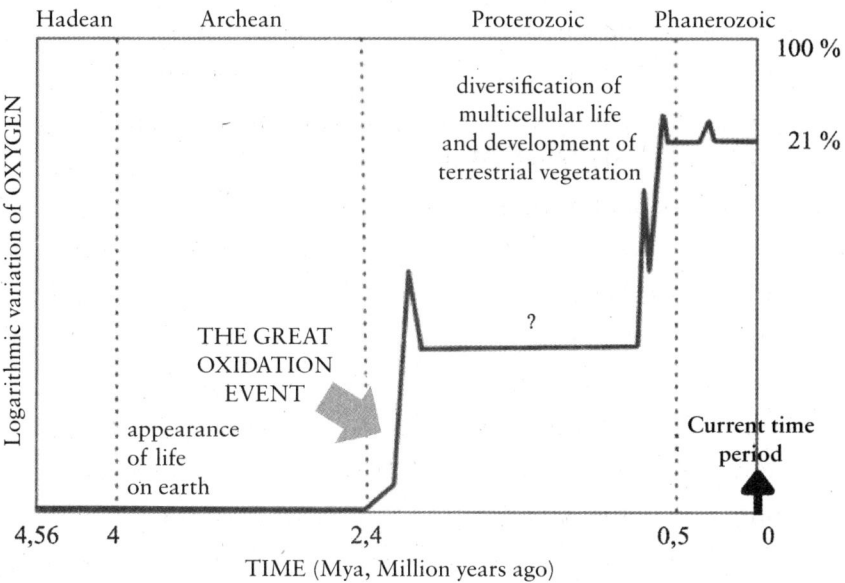

Oxygenation of the atmosphere through geological time:
impact of the Great Oxidation

Marine viruses, killer robots that enabled us to exist

Up to 100 billion viruses coexist in one litre of seawater.[21] Once again, what we see as empty is in fact a thick germ soup. And we swallow a decent helping of them whenever we drink a glass of water. But what exactly is a virus? It is a kind of parasite living inside a cell, and a single cell can contain up to 150 viruses. Marine viruses are believed to be the smallest, most abundant and most diverse organisms in the ocean. They therefore largely dominate this ecosystem. If we were to string all these viruses together like a pearl necklace, they would stretch for about 10 million light years, or about a hundred times the diameter of our galaxy.[22] Despite their astronomical quantity, viruses represent only 1 per cent of the earth's biomass (still equivalent to 75 million blue whales).

Despite their tiny size – a few hundred-millionths of a centimetre – marine viruses come in all sorts of incredible shapes. Some are icosahedral (twenty-sided), others are filamentary or shaped like multicoloured flying saucers. But the most surprisingly shaped are certainly the robotlike 'bacteriophages', which attack bacteria. These machine-like creatures, programmed to transmit infections, have a diamond-shaped head, a long body that ends in a tail for injecting their DNA, and six mechanical-looking legs. These viruses resemble some kind of metal spider from a science fiction movie. When they encounter a target, the attack is methodical: anchoring, injection, replication and explosion of the host cell.

Viruses are so ancient and genetically different that a common viral ancestor is unlikely. They would therefore have preceded the birth of LUCA, at the dawn of life on earth. At various stages they brought about radical innovations in the evolution of species. Some may have come from space. In any case, viruses have played a decisive role in the history of evolution. Through the ages, they have allowed genes to circulate within ecosystems. Moving from one organism to another to infect them, viruses carry a few genes with them. This transmission is essential in host cell mutations, helping them to acquire new abilities. These mutations allowed species to evolve more quickly. Here, we are talking about 'horizontal gene transfer', as opposed to 'vertical transfer', which takes place between generations (from parents to children).

Planktonic viruses play a fundamental role in the balance of life on our planet by moderating the development of plankton populations. They do this by annihilating a large number of them. Every day, 30 per cent of the world's photosynthetic plankton is destroyed by viruses! In the deep sea, nearly 80 per cent of plankton is contaminated and dies. Without these viruses, these plankton blooms could be so vast that they would cause a total planetary collapse. Some cover hundreds of kilometres of water surface and are so vast that they are visible from space. Yet, almost the entirety of this dizzying quantity of microscopic organisms can disappear within a few days following a virus attack.

The rate at which viruses reproduce is phenomenal. In optimal conditions, they can quadruple their population in a matter of minutes. These proliferations are generally positive for us because they eradicate a large number of pathogens. If viruses disappeared, we would not be able to survive. The infection of a single-celled plankton causes its death in the very short term, and therefore the death of millions of marine organisms which sink to the bottom of the oceans. There, in the underwater depths, their carcasses, composed of carbon molecules, sequester tonnes of CO_2, thus limiting atmospheric warming.

Unfortunately, in recent years we have seen a decline in the number of viruses in the upper layer of the ocean. As they are sensitive to solar radiation – particularly UV rays which degrade these organisms – virus populations are decreasing. And the impact of these populations on marine ecosystems is immense. Studies into the subject are also seriously lacking, which means this factor is not always taken into consideration in the mathematical models used to predict the consequences of climate change.

Moreover, while the planktonic viruses that regulate microbial communities are the fuel of the oceanic system and major biogeochemical cycles, some less important marine viruses also risk infecting animals. Those involved in aquaculture know something about this. Given our ignorance about marine viruses, the solutions are limited. First and foremost, upstream biosecurity is essential to ensure that drainage basins are not contaminated by viruses from household and industrial waste. As always, prevention at sea begins on the land.

We have recently seen that the development of new viruses can decimate populations of dolphins, otters or seals. In Hawaii, this phenomenon has led to the development of vaccines to save these populations from extinction. There are no strictly marine viruses currently known of that appear to be dangerous to humans.

The discovery of the first marine virus dates back to 1955, and they only really began to be studied seriously with the democratization of genomics in the 2000s. Over 99 per cent of marine viruses have been discovered in the last twenty years. There is cause for concern about the extent of our ignorance on a topic that is so fundamental for all ecosystems. We can also see it as an incredible field of exploration opening up for our future marine biologists, who now have the right tools to understand this invisible world.

Long before the Western world, the Polynesians discovered plankton (and because of that, the Americas)

The colonization of the Pacific by the Austronesians – and later by their Polynesian descendants – is one of the most amazing feats of early human migrations. The Pacific islands were so remote that they were among the last to be colonized by *Homo sapiens*. New Zealand (Aotearoa) did not see its first humans disembark until the thirteenth century.

At that time, reaching those islands meant weeks or months of navigation without any modern tools or maps, and with only an oral tradition and a great deal of bravery, because finding land at the end of a voyage was never guaranteed. Perhaps there would be no end to the dark blue horizon. Or perhaps it would lead to a huge trench populated by monsters, as our European ancestors believed?

It took not only courage, but also an incredible knowledge of the ocean for these populations to sail thousands of kilometres in search of hypothetical lands, on simple carved tree trunks.

That's why, long before us, these populations were able to deduce the existence of plankton and use it to navigate and feed themselves during their incredible expeditions! [23]

It all began nearly 3,000 years ago, the time when the great Greek navigators confined themselves to the 'small'

Mare Nostrum (Mediterranean), while the Austronesian peoples set sail on their pirogues from Taiwan to conquer the Pacific, and would reach the Philippines, Indonesia and Melanesia.

It should be pointed out that Australia was already inhabited by Aboriginal people who arrived there when Australia, New Guinea and part of Indonesia formed the vast continent of Sahul. A delta separating this continent from its Asian neighbour was crossed by rudimentary boats almost 40,000 years ago. [N]But for our Polynesian sailors, the challenge was much greater, as they had thousands of kilometres to cover!

Between 1600 and 800 BC, these brave men reached Vanuatu, New Caledonia, Fiji, Tonga and Samoa (Lapita culture). No human beings and almost no animals had ever inhabited these lands before, and the only things dotting the beaches were a few coconut trees that had grown from nuts drifting in the ocean.

From these ancient times, their sacred songs (*kumulipo*) recount the initial creation in the darkness of the oceans, life being born in the *pō* (night, fertile chaos) and evolving through successive generations.[24] Plankton (*ka uluwehi o ke kai*) is born first, followed in a very precise order by corals and algae, marine invertebrates, fish, birds, land plants, animals and, finally, humans.

This was almost 2,000 years before Darwin and his theory of evolution, but in essence everything was already there. These Polynesian peoples designed the foundations of a modern biology filled with respect for Nature and all that has preceded us, in this chain of creation.

It should be noted that for Polynesians, human beings come last, after the rest of the living world. This is very different from Western religious narratives that centre the creation of man. As a result, Polynesians are more integrated with the natural environment, filled with a respect and gratitude often lacking in Western civilizations.

Incredibly, here too we find this idea of plankton more than 1,500 years before the microscope was invented, and before the Europeans discovered the invisible inhabitants of the ocean.

The Polynesians knew how to make the most of these invisible lifeforms in their expeditions. From the first millennium onwards, they had to go even further afield to settle in Tahiti, Hawaii, then Rapa Nui and finally New Zealand. To navigate such long distances in a simple sailing canoe (*va'a taie*), two things were essential: finding their bearings, and finding food!

We now know that the Polynesians used the stars, clouds and birds to find their way. What is less well known is they also relied heavily on plankton.

At night, the luminescent plankton helped them to find their way, and to calculate the speed of their boat. They measured the time it took for the dots of light to travel from one end of the canoe to the other.

In the daytime, plankton gave texture, colour and even smell to the ocean, which formed the huge map on which they could find their bearings in this great emptiness. These nuances indicated the proximity of land but also, and above all, the presence of fish. These peoples understood that this invisible resource was the foundation of the food chain.

Wherever it was found, so was plenty of food. An opaque blue ocean indicated an oceanic desert, while a light blue or green hue was a sign of abundant fish and plenty of birds to hunt.[25]

This knowledge is what enabled the Polynesians to reach the Americas as early as the twelfth century, three centuries before Christopher Columbus and his huge modern caravels. Crossing the Pacific, however, is a much greater challenge than crossing the Atlantic. Even though there were no permanent settlements there at the time, DNA traces of Native American populations and the ancestral presence of sweet potatoes in Polynesia (which originated in South America, where they bear the same name) attest to the existence of exchanges between Americans and Polynesians long before the arrival of the Europeans.

Such feats were made possible by integrating plankton into a navigational ecology passed down orally from generation to generation.

Once they were settled on land, these Pacific peoples knew how to adapt their periods of fishing, the areas they exploited and their prohibitions (*rahui*)[26] to optimize their activity in a way that today we would describe as an ecosystemic approach. These practices were based on an understanding of ocean cycles, which we now know are clearly linked to the dynamics of plankton.

In many Polynesian traditions, the ocean is a living being, structured by invisible forces. The world is seen as a continuum between the visible and the invisible, the human and the non-human. This brings us closer to an ecological view of the Earth that is similar to the Gaia Theory which emerged in the twentieth century and which we will return to later.

The navigators of the Pacific read the patterns of the ocean the way we read computer code today. Their empirical system tracked plankton blooms as biological landmarks, transforming microscopic life into macroscopic navigational data, millennia before our instruments confirmed its accuracy. An invisible but fundamental form of life exists everywhere and is still the source of all marine and terrestrial fertility. Long before us, the inhabitants of the Pacific realized the importance of this...

We're bacterial beings, descended from archaea!

Subconsciously, we often associate bacteria with pathogens, responsible for illness and disease, even though only a tiny fraction of them are. In reality, it is the imbalance between different types of bacteria that poses a danger to our health. These organisms are, above all, the basis of all life on earth. The great American evolutionary biologist Lynn Margulis spoke about it this way in 1987: 'When we look at Planet Earth from a distance and across its entire history, it is a planet of bacteria, viruses and microorganisms. This is the essence of its history, of its innovation; this is its deep structure. We ourselves are – like all animals and plants – ultra-complex assemblages of bacteria; those microscopic beings form the continuous basis of life on earth'.[27] Bacteria are actually 4 billion years old, while land plants and animals are infants in comparison, at less than half a billion years old.

Within the biomass – excluding plants – bacteria represent 13 per cent of living things, i.e. six times more than fungi and thirty times more than animals. In the ocean, bacteria represent almost a quarter of the total biomass and most of them are plankton.

Bacteria have a crucial and little-known role in the nitrogen cycle. In order to grow, plants and algae need mineral nitrogen in the form of nitrates or ammonia, for synthesis of their proteins and other compounds needed to survive. In this virtuous cycle, when plants and animals decompose, they release ammonia into the soil or oceans. This ammonia is then transformed by bacteria into nitrate to feed plants, algae and photosynthetic plankton.

At the evolutionary level, bacteria will also cooperate in countless ways, contributing their potential to new organisms just as they have done throughout history. We will say more on this later on.

The adaptive capacities of bacteria seem limitless. Since their origins, they've been able to reproduce and develop under any conditions. While blue bacteria invented photosynthesis, others became heterotrophs (they break down organic matter to feed themselves) or mixotrophs (they practice both photosynthesis and heterotrophy). Within ecosystems, they recycle dead organisms to release nutrients into the water and make them available to other life forms. Precisely as they do inside our gut.

The interactions between bacteria and their environment are still largely unknown, but could be even more extraordinary than we think. At the human level alone, it is now proven that the microbiome is intimately linked to our brain function and our unconscious decisions. Ultimately, our physical body may be nothing more than a container for these colonies of bacteria.

Archaea are even less well known, yet also play a greatly underestimated role on our planet. Mistakenly referred to for a long time as 'archaebacteria' (archaic bacteria), they both share the characteristics of being unicellular and lacking a nucleus, with structures that appear quite similar at first glance. However, these similarities are only superficial. It was not until 1977, and the work of Carl Woese, that scientists clearly differentiated archaea from bacteria. In terms of their chemical composition, molecular biology and genes, bacteria are as different from archaea as they are from multicellular organisms like us.

For a long time, science limited their presence to 'extreme' areas, characterized by very high temperatures, high levels of acidity and low oxygen levels. We now know that archaea feed on sulphur and iron. On the other hand, they don't like oxygen and took refuge in the depths of the ocean during the Great Oxidation. They form strange alliances with all kinds of organisms, and sometimes even sponges. Like bacteria, they are also present in our stomachs, forming part of our microbiome.

In all environments where oxygen is absent – in rice fields, landfills, peat bogs and marshes – archaea are at work creating methane. Even the smallest cluster that is sealed off from oxygen, and if possible from light, allows them to proliferate. They also produce methane in our stomachs, and especially in those of ruminants. Because of their multi-chambered stomach, cattle emit this gas which is 80 times more destructive to our climate than carbon. Methane accounts for more than 20 per cent of our greenhouse gas emissions. The climate impact of cattle farming on the planet is therefore largely linked to these tiny archaea hidden in their intestines.

Through this release of methane into the atmosphere, the proliferation of archaea has always been the source of significant global warming, and even heatwave periods with temperatures rising above 100 degrees centigrade. But it is thanks to them that our planet was able to emerge from terrible ice ages, allowing our current ecosystems to develop. We now know that our human cells, like all animal and plant cells, are descended mainly from archaea.

Bacteria and archaea are the two pillars and the oldest representatives of life on our planet. Today, they form two of the three major domains of the living world. Of the millions of life forms that tried to exist during the first billion years of earth, only these two types of organisms managed to find their way from LUCA to us.

Joining forces to survive,
or the magic of 'symbiosis'

Ever since the first bacteria left the seabed to become plankton, drifting in the ocean to encounter other organisms, two factors have been key in evolution: time and cooperation. The timeline of these primitive organisms spans hundreds of millions, even billions of years. When it comes to cooperation, although we often view evolution as a competitive process, the greatest biological advances are more often the result of alliances, symbiosis[28] and close collaboration between two organisms that exchange energy and resources in order to adapt, together, to their environment.

Take corals, for example: these animals, like a multitude of invertebrates found on tropical reefs (such as molluscs, jellyfish, sponges), depend on a certain type of plankton, zooxanthellae. These provide the corals with carbon-based energy (sugar, lipids) through photosynthesis, and in exchange the corals release the nutrients the zooxanthellae needs. The term 'holobiont' is used to refer to an assemblage of organisms that collaborate to form a single organism. Humans form a holobiont with our microbiota.

There is also a special form of symbiosis called endosymbiosis, in which a host cell incorporates another cell into it, in exchange for resources and protection. Many scientists even believe that historically, the symbiotic relationship (alliance) preceded the predatory relationship (acquisition). So, long before the invention of sexuality, it was already clear to our ancestors that it was more effective to work together than to devour each other.

The most striking example of this is our own cells. Two billion years ago, planktonic bacteria capable of providing energy were integrated into other cells to accelerate progress. These bacteria became our mitochondria. Humans did not invent the race for progress, or the race for energy. Although wood, hydrocarbons or nuclear power have been part of our recent civilizations, energy has always been at the heart of life. For the first cells, the sun was enough. But life quickly became more complex and the process had to be improved to speed things up. Mitochondria, champions of energy production, were quickly identified as an interesting resource. Ingesting the same amount of sugar, bacteria are capable of producing eighteen times more energy compared to a normal cell. Our cells' ancestors therefore integrated these mitochondria through endosymbiosis, to benefit from their energy. Having these mini powerplants inside the cells themselves was the only way to achieve the output necessary for an organism as complex and energy-intensive as an animal. Throughout evolution, we also acquired part of their genome in our nucleus. A molecular dialogue has allowed us to domesticate these bacteria.

Even today, these symbiotic mitochondria – present in their thousands in each of our cells – allow us to live, breathe, move and think. Inside us, these bacteria retain their own DNA. We are all therefore composed of a mosaic of organisms with their own biological history. And each cell is a collection of holobionts that evolved in planktonic form in the ocean, until they became what we are today: plural beings.

In addition to the acquisition of mitochondria through endosymbiosis, plants and microalgae have benefited from other symbioses too. By integrating photosynthetic cyano-bacteria, algae, the ancestors of our plants, acquired the ability to transform solar energy into organic matter. These bacteria, which became chloroplasts, allowed land plants to thrive and become the main producers of oxygen on earth. It is thanks to this symbiosis that we have the forests, the meadows and even the oxygen we breathe today. The whole of evolution is nothing more than a succession of endosymbiosis that could inspire modern-day followers of transhumanism.

Everything we are today started in the form of plankton: tiny organisms floating in the ocean, establishing the first major alliances of the living world. All of evolution has been built on these collaborations. And perhaps the best is yet to come. The possibilities are beyond our imagination.

Perhaps we might become photosynthetic beings in the future. This would limit our impact on the environment since we would no longer need to feed ourselves. This technique is already used by some animals: corals of course, but also terrestrial animals such as the Roscoff worm (*Symsagittifera roscoffensis*), the oriental hornet, the Elysia sea slug or the eggs of the spotted salamander. They all ingest microalgae, but instead of digesting them, they integrate them into their organism to become partly photosynthetic. The Chilean company Symbiox already offers microalgae-based dressings to oxygenate wounds.[29] The photosynthetic activity of plankton accelerates the healing of damaged tissues. The way forward is already clear!

In our modern world that is undergoing an energy and climate crisis, the hypothesis of a symbiosis between humans and photosynthetic organisms might one day lead to a completely autonomous 'Homo photosyntheticus': a mixotrophic being, capable of capturing the sun's energy and producing its own oxygen. That would certainly make long-distance space travel and the colonization of other planets a lot simpler. Acquiring chloroplasts in this way would be consistent with the way our plankton ancestors acquired mitochondria a long time ago.

Ultimately, though, what matters most about these multiple partnerships is this: life did not conquer the earth through force and combat, but by weaving a network of heterogeneous sharing. We are a long way from the 'survival of the fittest' idea popularized by proponents of Darwin. The greatest lesson that plankton can offer us – after billions of years of evolution – is to show us that for survival and innovation, *cooperation* is always more powerful than *competition*. This principle should inspire us to think about what our future is going to look like. Especially since, in the case of plankton as well, cooperation can be fragile and very often evolves into parasitism when one of the partners 'cheats'...

Plankton, the painter of the ocean

We tend to describe the sea as blue, but in some parts of the world it can take on red, green, yellow or milky-white hues. This varied colour palette is down to the presence of plankton. Omnipresent in the oceans and sometimes in freshwater, plankton significantly affects the chromatic variation of our aquatic ecosystems. In spring, under the effect of the sun, the high concentration of phytoplankton can make the sea turn green.[30] Certain red algae blooms cause terrifying scarlet tides, which are as toxic to marine life as they are to coastal inhabitants. In the Bible, we see the Nile turn to a river of blood (Exodus 7-21): 'The fish in the Nile died, and the river smelled so bad that the Egyptians could not drink its water. Blood was everywhere in Egypt.' In reality, this was almost certainly a bloom of scarlet plankton, or 'red tide'. As for the Red Sea – located between Arabia and Africa – the name does not come from the colour of the water, since it is blue, but more likely from episodes of reddish plankton blooms.

Phytoplankton (or microalgae) naturally contain carotenoids with orange pigments that help protect themselves against UV and oxidative stress. So *Dunaliella salina*, which is initially green, turns orange or red when exposed to very salty waters, strong light or a lack of nutrients. This is what sometimes gives salt marshes, or the Dead Sea, for example, a pinkish hue. In addition, certain crustaceans, shrimp or krill also feed on *Dunaliella salina*. The colour pigments are then passed throughout the entire food chain, to salmon, trout, aquatic fish and flamingos. While male flamingos like to strut around and show off their beautiful plumage, females use this antioxidant pigment to as a deterrent, to protect their eggs. In other latitudes, zooxanthellae colour corals with hues of yellow, pink, green or red. Viewed from the sky, these immense patches of corals can sometimes cover thousands of square kilometres and form surprising shapes.

This colouration phenomenon is not limited to the ocean. In freshwater, the aptly named *Haematococcus pluvialis* is known for causing the rains of blood that terrorized our ancestors. They saw this as yet another divine punishment.[31] Similarly, blood snow (also known as red snow, watermelon snow or glacier blood) is the chromatic consequence of another microalga. Aristotle was already observing it on Mount Olympus in winter, 2,500 years ago. He imagined hairy larvae. Much later, we thought it might be mushrooms. It wasn't until the end of the nineteenth century that it was understood that all of this was down to micro-algae. And only in 2019 was the exact culprit identified: the genus *Sanguina*, a green microalga that covers itself with reddish antioxidants to protect itself from the excessive light reflected by the snow and from extreme cold. These microscopic algae are found at up to 3,000 metres altitude in the Alps, and in Greenland the ice sheet is dressed in black or pink.

We still don't fully understand what causes this phenomenon. How do the algae end up in glaciers? Where do they go when the ice melts? What we do know is that these phytoplankton prefer waters that are rich in phosphate. Their proliferation is also an indicator of water pollution by insecticides, pesticides, heavy metals, etc. On the one hand, they play a purifying role by absorbing excess mineral and metallic nutrients; on the other hand, they multiply in such proportions that they risk causing ecosystem imbalances. This also alters the way our seas interact with the sun's rays, contributing to the formation of storms and cyclones. By changing the colour of the ice, microalgae alter its capacity to reflect the sun's light and heat.[32] This phenomenon therefore accelerates glacier melt; they are currently melting at a rate that cannot be explained by rising temperatures alone. A tragic fate for these microalgae, which are unable to survive in other conditions. The glacier blood they create is, in a way, their own. The widespread occurrence of these planktonic blooms is extremely worrying. They pose a danger to both these algae and to the glaciers, but also to the planet as a whole.

Magical plankton that illuminate the waves

Plankton is not only the painter of the ocean, but its lighting designer as well. Few people know this, but 76 per cent of marine species that inhabit the water column (pelagic) are bioluminescent.[33] In the moonlight, they can illuminate the ocean with a thousand glittering rays, gliding over the waves like a psychedelic blanket of stars. Jellyfish, over 1,500 species of fish and even some sharks have inherited this dazzling talent. They all inherited it from their planktonic ancestors.

Depending on the species, the glowing trails of light can be red, orange, blue or green. Most of these phenomena occur at the surface, but this stunning magic can be observed at a depth of down to 4,000 metres.

Bioluminescence capabilities obviously vary from one species to another. In this sector, certain plankton stand out as stars of the show. Their cold light requires a lot of energy to create, and produces almost no heat. It can be used to ward off predators or, conversely, to attract prey that believe they are heading towards the sun. Paradoxically, some species even use light as a means of hiding... Some predators living in the depths spot their victims thanks to the shadow they cast as they pass over them. Producing a halo of light will therefore successfully fool those hunters. For others, the light will be used to communicate or simply to illuminate their path through the abyss. Finally, some resort to flashing, in the hope of attracting a sexual partner to reproduce with. Nightclubs didn't invent anything new.

In the ocean, these phenomena can offer amazing spectacles when in proliferation. Anyone who has experienced the magical ballet of bioluminescent plankton around their boat will remember it for the rest of their life. Movements in the water usually accentuate the production of these lights that dance on the horizon. On beaches, these efflorescences create shimmering swells that send out blue or green beams and then break on the rocks, shattering into a thousand fluorescent fragments.

Although bioluminescence is common in underwater life, not everyone will get to see these waves of light. To see it, you need a calm sea, warm water, a new moon so the sky is not too bright, very little wind and, above all, a large quantity of nutrients rising from the sediments to the surface thanks to the currents.

It is therefore rare to see bioluminescence in Europe. You have to travel to Puerto Rico, Japan, Australia or California to witness it. On those beaches, at night you will find surfers playfully tracing trails of light with their boards among the phosphorescent waves. It is nothing new though: Jules Verne mentions it in *Twenty Thousand Leagues Under the Sea* and some fishing boats have always known to follow these halos of light in the ocean. Because where there is sparkling plankton, you also find the fish that hunt it.

There are few studies relating to these bioluminescent plankton. There was a recent resurgence of interest among Australian military personnel when they realized that the trajectory of their precious submarines became visible to the naked eye as soon as they entered a cloud of luminescent plankton… or one might even say diabolical plankton, its light produced by a molecule aptly named 'luciferin'.[34]

In the Russian Arctic, luminescent snow has even been observed by researchers in some places. They witnessed a strange sight, like a blue aurora borealis spreading across the pristine ice floe. Each of their footprints created an even more intense path of light, and snowballs sparkled brightly between their hands. It seems that these polar Christmas lights are the work of microscopic crustaceans that the winds and tides have carried to the ice floes. These plankton – the very abundant copepods – glow here in an ultimate icy breath.

Bioluminescence has evolved at least forty times independently in the living world, according to scientists.[35] The mechanism is a time-honoured classic. Even in humans, vestiges of it remain. Certain biochemical reactions generate very small amounts of light, usually too faint to be seen with the naked eye. Fireflies and glow-worms are the only land animals to have actually retained these properties from their planktonic ancestors.

Protists, the dark matter of life descended from Loki, Odin and Thor

Drifting in the ocean in the form of plankton, bacteria and archaea have founded and shaped life for 4 billion years. But in terms of composition, our cell types are profoundly different from archaea and bacteria. These cells seem to have appeared on earth just over a billion years ago. But do they have a link with archaea and bacteria?

This seemed likely but until very recently this evolutionary leap remained incomprehensible. How did we go from a single microbial cell with a single chromosome to specialized cells with more than twenty chromosomes, capable of grouping together to form complex and highly sophisticated entities? Part of the answer to this was found in one of the most inhospitable places on the planet. A raging furnace hundreds of metres deep in the icy waters north of Scandinavia: the now famous 'Loki's Castle'.

It all started in 2008, with a team from the University of Bergen going on a diving expedition to the 73rd parallel. They discovered incredible hidden hydrothermal vents between Greenland and Norway. Black smokers!

These vents are located at depths of over 2,500 metres. Down there, the pressure is equivalent to the weight of 600 elephants standing on your hand. The chimneys spew out sulphurous water that comes from the bowels of the earth and can reach temperatures of 350°C. Light doesn't make it down this deep, while the high iron and manganese content gives the water a very particular black colour; hence the name. These newly discovered black smokers are the northernmost on our planet. Five bottomless pits erupting into an eternal night at the bottom of the sea. It seems as if nothing is able to survive there. The place is called 'Loki's Castle' in reference to the shadowy Norse god associated with fire and chaos. In 2015, finally having the equipment capable of descending to these depths to study life down there, researchers from Uppsala University launched a scientific project. As expected, they found archaea, but not just any archaea. A very special kind of mutant archaea, exhibiting profound modifications never observed before. These archaea have surprising genetic and metabolic similarities with plant or animal cells, and therefore human cells.

The scientific community quickly understood that these organisms were the missing link in cellular history.[36] It's somewhat like finally discovering the mutant hominid that directly links monkeys to the human species. Within its furnace, Loki's Castle contains the link between very ancient archaea and the modern multicellular biodiversity to which we belong. The first archaea found there were named Loki in reference to the black fiery 'castle' where they were discovered. Later, other similar types were discovered in Slovenia, North Carolina and Japan. These were identified as 'cousins' of Loki's archetypes and were therefore named Odin, Thor and even Heimdall. The group of archaea was thus named the 'Asgard archaea' in reference to the realm of the Norse gods.

As research went on, it was discovered that their metabolism did indeed exhibit preliminary elements for the formation of a compartment inside their cell, i.e. a cell nucleus.

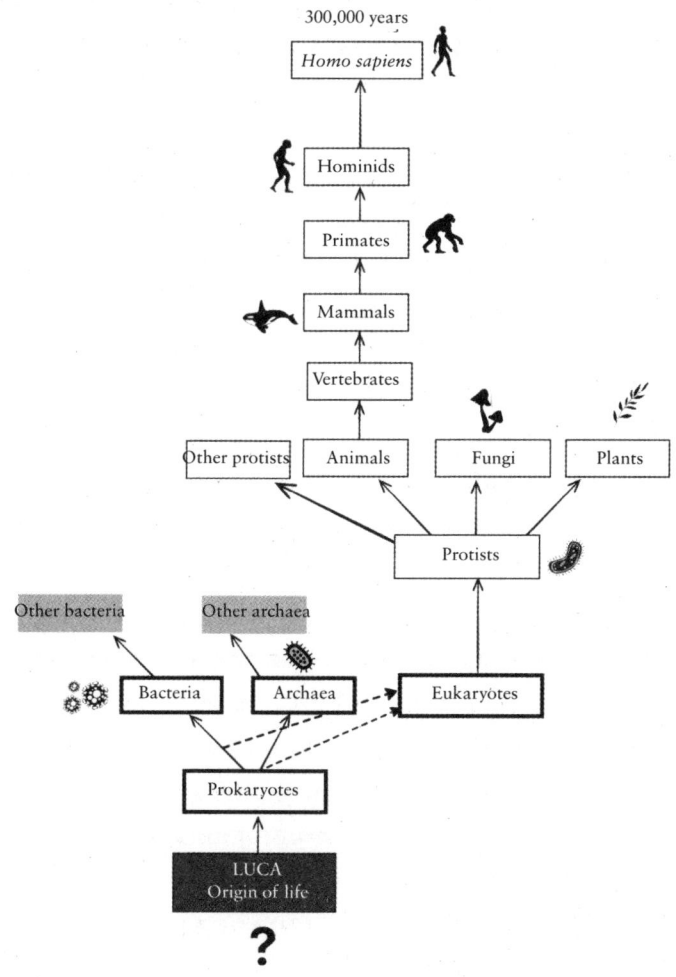

(Very partial) classification tree of Life

This invention represents a major schism in the history of evolution and marks the beginning of a sudden acceleration.

At the time, the only living cells were prokaryotes which, by definition, have only one chromosome and lack a nucleus, just as bacteria and archaea still do today. The formation of a nucleus would allow a third type of cell to emerge: eukaryotes.

The nucleus of these newcomers allowed for better protection of acquired genetic material, and for increased diversity by enclosing multiple chromosomes. In the early days, these eukaryotes remained unicellular and made up the still largely unknown group of 'protists'.

Protists, often overlooked, have played a central role in the history and balance of life on our planet for almost two billion years. Today, eukaryotes are divided into four groups: plants, animals, fungi and... protists.

Of the four kingdoms, protists have by far the greatest diversity and are the origin of the other three.

Protists make up over a third of the mass of living organisms in the ocean. At 2 gigatonnes in total, their weight is more than that of all marine animals combined (dolphins, whales, turtles and other fish). Moreover, from oxygen production to carbon sequestration – including the nitrate, phosphate and freshwater cycles – their importance for the balance of life on earth is far more significant than all the other groups. Yet, they aren't mentioned in any schoolbooks and, with the exception of the group of protists that make up phytoplankton, very few scientists are interested in them. This collective omission has earned them the nickname, 'dark matter of life'.

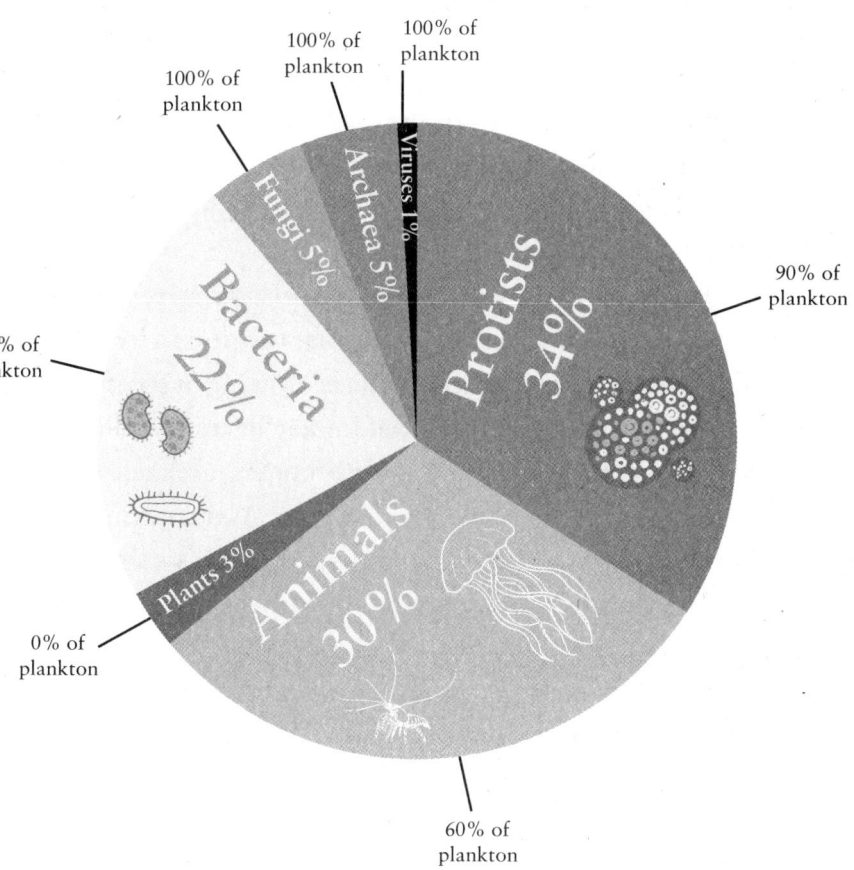

100% of plankton
100% of plankton
100% of plankton
100% of plankton
90% of plankton
Protists 34%
Viruses 1%
Archaea 5%
Fungi 5%
Bacteria 22%
% of
nkton
0% of plankton
Plants 3%
Animals 30%
60% of plankton

Contributions of plankton to total ocean biomass

Drawing inspiration from similar bacterial techniques, these unicellular eukaryotes also contributed another major invention to the history of life: sexuality.[37] Through sexual reproduction, genetic mixing accelerated, generating increasingly complex and adaptable organisms.

The Asgard archaea of Loki's Castle are therefore the vestiges of a decisive break that led to the emergence of eukaryotes. All living things on earth today are either bacteria, archaea or eukaryotes. From oak trees to humans, including sea sponges and amoebas, we are all are descended from protists, and therefore from the Asgard archaea. Another crucial innovation – multicellularity – was still needed for these hitherto unicellular eukaryotes to give life to all the organisms we see today.

But this burning castle in the dark depths of the icy waters off Greenland reminds us, above all, that we are much closer in evolutionary terms to those archaea evolving in terrifying conditions than we are to more familiar bacteria. What's more, unlike bacteria, there are no known archaea that are pathogenic to humans. We're family! And contrary to what the Romans told my ancestors the Gauls, we are all descended from Asgard, not from Olympus.

Diatoms: the jewels of the ocean fundamental to aquatic life

Long before their ecological importance came to be recognized, diatoms captivated us with their beauty. Observed under a microscope, these myriads of aquatic crystals resemble delicate lacework carved by an artist. The colonies of glass-coated diatoms form undulating, diamond-like forms. From lenses to elliptical shapes, from polygons to rosettes, these 'jewels of the ocean' adorn themselves with fantastical silica shells, sculpting worlds in the heart of the ocean. Their exoskeletons, called frustules, are made of a thin layer of glass that protects them from predators while allowing the light necessary for their photosynthesis to pass through. Small pores also facilitate the exchange of nutrients with the outside. Darwin himself acknowledged, in *On the Origin of Species*, that 'few objects are more beautiful than the minute siliceous cases of the diatomaceae'.[38]

Each of these incredible glass masterpieces is infinitely recyclable and resistant to almost anything. In the laboratory, only the force of ultrasound can shatter their stained glass. But the most amazing thing about them is not their almost supernatural beauty. Their role in the planetary system is even more astonishing. These elegant plankton are responsible for 25 per cent of primary matter production,

20 per cent of oxygen creation and 20 per cent of carbon sequestration.[39] Diatoms alone outnumber all tropical rainforests. We owe them one fifth of the renewed oxygen that we breathe every day.[40]

Appearing around 200 million years ago, diatoms laid the foundations for terrestrial life as we know it. They continue to be its pillars, and are present in every single drop of fresh or saltwater. They cover aquatic soils and travel with the currents, preferring cold waters. The greatest concentration of these plankton is found beneath the ice sheet of the polar regions.

Whenever you step in a puddle, you crush millions of these diatoms. Waterfalls, damp walls, gutters... these microscopic diamonds are everywhere. In colonies, they sometimes appear as brown or yellow trails along a quay, in a pond or on a beach. Every litre of water contains about ten million of them. They represent the most widespread vegetal group on earth and constitute the vast majority of plankton. The number of diatom species remains uncertain, ranging from 30,000 to over one million. They were not discovered until 1702, and it then took another century to understand that they were not animals. Primarily photosynthetic, they are actually related to plants. They are also called microalgae, but some of them are mixotrophic, capable of also feeding on organic matter.

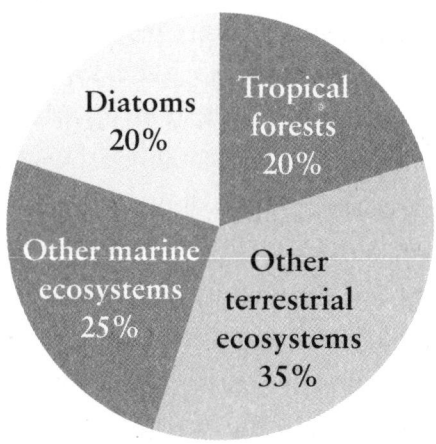

Diatoms
20%

Tropical
forests
20%

Other marine
ecosystems
25%

Other
terrestrial
ecosystems
35%

Photosynthesis on a global scale (oxygen production)

These ocean magicians have plenty of other tricks up their sleeve. They know how to create glass in the cold and they divide at breakneck speed. Their name comes from the Greek *diatomos* meaning 'cut in two'. A single cell can produce millions of individuals in a matter of days.[41] However, this process limits their growth: by repeatedly dividing, they decrease in size. After a while, they must adopt a surprising strategy to recreate a large diatom: they alternate between vegetative (asexual) and sexual reproduction! This is a unique phenomenon in the living world. When they reach a critical size, they release gametes that produce new, genetically distinct individuals, capable of restoring biodiversity while reconstructing a large diatom that can be divided. This new diatom could be up to ten times larger than its parents. Anyone raising a pre-teen should be glad that this characteristic was not passed on to humans.

In the ocean's food chain, diatoms feed zooplankton,

which digest them by releasing their frustules in the form of faecal pellets. This waste, weighed down by silica, sinks to the bottom of the oceans and forms what is poetically called 'marine snow'.[42] After travelling for weeks, this 'snow' traps carbon in ocean or lake sediments for millions of years.

Over time, diatoms have filled our atmosphere with oxygen while removing carbon, cooling our climate. Their deposits have also covered the earth's layers with a sedimentary rock, diatomite. In Mexico, this reaches a thickness of up to 1,000 metres. In France, part of the Massif Central region is made up of this diatomite, which is transported out of the water by geological movements.

Diatoms also have many industrial applications: they are used in road markings (as they reflect light), moisturising face masks, medicines, paper, tyres, cat litter, dynamite, matches, filters, organic insecticides and various construction materials.

Above all, diatoms have enabled the emergence of current life forms by sequestering carbon, releasing oxygen, but also by absorbing silica to significantly alter the chemistry of the ocean. They are now used as an indicator to measure 'water quality'.[43] Unfortunately, although they are resilient, diatoms are very sensitive to their environment. Climate change, residues from our land-based pollution and aquatic infrastructure will all have a strong impact on them and encourage the proliferation of other planktonic communities that are often less beneficial to our ecosystems. We urgently need to gain an understanding of how to preserve these little-known jewels, which hold in their frustules (their external walls) the balance of life as we know it today.

Coccolithophores: cathedrals on the cliffs of Étretat

Although diatoms make up the vast majority of photo-synthetic plankton, they are definitely not the only representatives of this diverse and prolific universe. Among other notable players, coccolithophores are part of microalgae (and therefore plants) because they carry out photosynthesis. They gather in colonies that create a white foam on the surface of the sea.

Appearing several hundred million years ago, coccolithophores were some of the major contributors to oxygen production. But they are more well known for their ability to use calcium carbonate dissolved in the ocean to make their external skeleton: the coccolith. These microstructures, with their various magnificent shapes, resemble strange vessels, and clearly inspired the Death Star in *Star Wars*.

Once abandoned by their hosts, the splendid coccoliths also sank as 'marine snow' to the bottom of the oceans. As they settled, they sequestered carbon and created limestone. Over time, these layers transformed into chalk and friable soils that the evacuation of water or the movement of continents sometimes brought out into the open air. Thus, the Cretaceous period, or 'chalk age', owes its name to a huge proliferation of coccolithophores. Around 70 million years ago, the Paris Basin was still submerged by a shallow tropical sea where layers of limestone were deposited. So Paris is built on a pile of dead creatures!

Whenever we use white chalk on blackboards, we are unknowingly using the shells of these tiny beings. Likewise, the unique taste of champagne comes from these limestone deposits which form the basis of inimitable Champagne-Ardenne soil on which the grapes grow. Even more incredibly, Rouen Cathedral, the pyramids of Egypt and the pillars of the Hagia Sophia Basilica in Istanbul are all built from these coccoliths.

Their sediments also give the dunes of the Western Desert of Egypt their particular white hue. Some of these layers of planktonic limestone were compressed and pushed out of the ocean by the tectonic upheavals of our planet. They created splendid cliffs such as those of Étretat or Dover. These remarkable walls of rock towering above the ocean are composed of 98 per cent of these planktonic remains.

Coccolithophore blooms sometimes cover entire seas and can reach the size of a country. These phenomena – which are visible from space – can influence the climate. Typically, having dense plankton at the surface reduces the temperature of the ocean, and therefore the risk of cyclones. But the immaculate white coccolithophores reflect sunlight and alter this balance.

The interactions between climate and plankton do not end there. Like other marine microorganisms, they produce dimethyl sulphide (DMS), a gas that plays a key role in the water cycle and cloud formation. This phenomenon contributes, among other things, to the distribution of precipitation on earth. Although they are located at the same latitude and both border the ocean, Brazil and Namibia have radically different climates: the prevailing winds blow westward and carry DMS to South America, favouring rains and lush forests. In doing so, they deprive Namibia of the marine gas that makes it rain, turning this country into the oldest desert in the world.

As well as terrestrial vegetation, marine life benefits from these rains which hold potassium, iron and silica. These nutrients enable crustaceans and shellfish to form their exoskeletons. In addition to influencing the climate, DMS gives the ocean its characteristic odour. This scent is used by certain seabirds such as albatrosses to locate areas rich in plankton, and therefore in fish.[44]

Finally, coccolithophores also have a significant impact on the surrounding biodiversity. Although their presence is mostly positive, just as with certain types of diatoms, coccolithophore proliferations can sometimes unbalance the ecosystem. For example, some cod larvae feed on microscopic zooplankton, which themselves ingest tiny plankton. When large coccolithophores take the place of these tiny plankton, small zooplankton, incapable of finding anything small enough to ingest, starve to death. The rest of the chain collapses and the cod disappear.

Coccolithophores have produced a lot of oxygen, but underwater life is still full of paradoxes for anyone who tries to apply exclusively terrestrial logic to it. For example, it is commonly said that the ocean provides half of the oxygen in our atmosphere, and we owe one in two breaths to the ocean. This regularly repeated statement has sparked a great deal of scientific debate because it suggests that without the ocean, we would no longer be able to breathe. Which is completely inaccurate.[45] Current oxygen reserves in the atmosphere represent an autonomy of approximately 1 or 2 million years for terrestrial life on earth.[46] Furthermore, almost all the oxygen currently produced by plankton is consumed by aquatic life. The amount that reaches the atmosphere and becomes available to us is therefore very small. Most of the new oxygen produced in the atmosphere comes from plants. The fact remains that a significant portion of the oxygen reserves accumulated in the air was indeed produced by plankton at a time when underwater life, being less prolific, consumed less of it. The oxygen produced by plankton throughout history has allowed the development of other land plants. So from a historical point of view, we owe all of the oxygen on the planet to the ocean.

Multicellularity: a biological big bang invents ageing

For reasons that are still a mystery to us, the creation of the nucleus and then the invention of sexuality provided a new pivotal moment in the history of living things. Instead of developing into colonies of similar cells, as archaea and bacteria did, new eukaryotic cells began to specialize and participate in a collective project that was as complex as it was ambitious. Life became multicellular! This divergence constituted a major turning point in the history of biological evolution, which some have compared to the Big Bang that created the universe.

The first multicellular organisms had cells that specialized in specific tasks: nutrition, mobility, formation of a shell, an epidermis or even a skeleton. Hundreds of millions of years later, these organisms developed muscles, a brain, and sensory organs. Multicellularity allowed living beings to become bigger, more robust, mobile and resistant, but the need for the cells to work collectively presented its own challenge. Multiplicity and collaboration inevitably cause tensions, misunderstandings and even conflicts. It quickly became apparent that, to avoid escalation, cells need to renew themselves. Hence the need for ageing and above all, death.

Programmed death, or apoptosis, is a key element of this

organization. Unlike unicellular organisms, which only died due to extreme and external events (predation, accidents, sudden changes in environmental conditions, etc.), the cells of multicellular organisms learned to renew themselves. This evolution encouraged cooperation between cells to maintain the integrity of the organism, while accepting the programmed end of certain cells. And so, as beings became more complex, ageing and cell death became the norm. Sometimes, certain cells malfunction, decline faster than others, or go rogue, no longer adhering to the collective endeavour. These antisocial behaviours are known today as 'cancerous tumours'.

The earliest examples of multicellularity date back more than 2 billion years, thanks to fossils found in Gabon.[47] These organisms, like red algae, came from symbioses between different types of cells. Multicellularity probably appeared twenty-five times throughout history, before becoming established around 1.3 billion years ago. And it took several hundred million years for more complex multicellular organisms, such as animals, plants and fungi, to emerge.

This explosion of multicellularity was followed by rapid evolution, with the appearance of new organs allowing adaptation to the environment. For example, the formation of fins and muscles has facilitated animal movement, their encounters and cooperation. Competition compelled them to innovate. New life forms emerged, resulting in the dramatic diversification of marine species.

Ecosystem destroyers and poisons used by the CIA: history's villains

Every story must have its villain. When it comes to plankton, this role is often wrongly attributed to dinoflagellates. As their name suggests, these protists have two flagella that spin to make them move up or down in the water. Hence their name, derived from the Ancient Greek δῖνος (dînos) meaning 'whirling' and the Latin flagellum, meaning 'whip'. Unlike diatoms, they can move independently. This advantage has helped to accelerate the evolutionary process of life on earth. And our sperm cells have retained this invention, even though they seem to have thought that a single flagellum was enough to find its way to an egg!

Some dinoflagellates use their flagella like lassos – Indiana Jones style – to capture the microalgae that constitute their diet. Capable of alternating between predation and photosynthesis, these mixoplankton adapt to their surroundings: if there is no light, they devour diatoms, and in the absence of available prey, they use the sun's energy. This is why they can be found even at great depths. They are essential for regulating the ecosystem. However, they have two disadvantages that tarnish their image slightly.

Firstly, some of them emit toxins. Of the millions of plankton species only a few are toxic, but the majority of those toxic species are dinoflagellates. The toxins from these protists accumulate in shellfish (oysters, mussels, scallops, etc.). They do not harm the shellfish but are dangerous for the humans who consume them, and can cause all sorts of ailments (diarrhoea, numbness, skin irritation, neurological disorders, even fatal paralysis). Saxitoxin, released by *Alexandrium catenella,* is much more toxic than cyanide or sarin gas, to the extent that it was the only marine toxin included in the list of 'chemical weapons' at the 1993 International Convention. The CIA even used it to make pills, which were distributed to its pilots so they could commit suicide in the event of enemy capture. The effects are irreversible, because there is no known antidote.

As early as 1799, Russian sailors reported the deaths of more than 100 men in one evening after they were poisoned by contaminated shellfish. In North America, they talk about 'Alaskan roulette', because seafood consumption can be so risky. This partly explains the tradition of only eating seafood during the months containing the letter 'R', as this toxic plankton is abundant in the summertime.

Dinoflagellates are also responsible for the red tides that wash up on beaches and destroy everything in their path. As with any poison, the dose is what makes it toxic. This type of plankton proliferates as a result of sunshine, human pollution and warming waters, to the detriment of other friendlier species such as diatoms and coccolithophores.

An imbalance can affect all the surrounding biodiversity, and may even threaten our own existence. Dinoflagellates do not have a bad reputation because of their worrying toxic properties. Their impact on the climate also raises concerns. Unlike diatoms and coccolithophores, their lightweight shells do not sink down to the deep seabed and sequester carbon. By remaining on the surface, they quickly release their carbon into the atmosphere.

However, climate balance depends on the ability of plankton to sequester carbon. If dinoflagellates were to become dominant, the ocean's capacity for carbon uptake could decrease, significantly exacerbating global warming. If we do not intervene to disrupt this mechanism, the ocean will eventually end up emitting more CO_2 than it absorbs, which would mean an irreversible disaster, regardless of any parallel efforts. Evolution will then transition to a new form of life, from which we will most likely be excluded.

However, there are solutions. The science of plankton is a still untouched horizon that has so much to offer us. We now have the technological tools we need to take action. But we must still choose to use them.

Copepods: the micro-factories
that feed the ocean

The copepod is undoubtedly, and quite unintentionally, the most well-known plankton in the world. This transparent microcrustacean was popularized by the character of Sheldon James Plankton, the tiny green villain in the animated TV series *SpongeBob SquarePants*. But this incredible group deserves recognition for many other reasons. The term 'copepods' includes more than 14,000 marine and terrestrial species, none of whom are larger than a millimetre in size. There are more of them in the ocean than the total number of all insects, combined, on earth – just think about that. Their biomass is one of the largest in the world and copepods are by far the most numerous of the planktonic animals.[48] In some areas, they make up to 80 per cent of zooplankton! Moreover, they adapt to all of the world's water types, from ice caps to tropical salt seas, from underground water tables to high mountain lakes, including rivers and salt marshes.

Some species are parasites and present a wide variety of morphologies. Their single, central eye has long earned them the nickname 'cyclops' or 'monocle'. More recently, the term copepods refers to the fact that they 'row with their feet'. In Greek, *'kope'* means 'oar' or 'paddle', and pod means 'foot'. They actually move using their antennae by doing little jumps underwater. Under the microscope, these little leaps are actually gigantic and amazingly fast, considering their tiny size. They are therefore the fastest animals in the world by a long way.[49, 50, 51] With top speeds of 120 centimetres per second, for organisms that are 3 millimetres long. That would be like a human moving at 2,700 km per hour, which is the speed of a latest generation fighter jet. Furthermore, the copepod is capable of lightning-fast accelerations, which for us would be equivalent to 55,000 times the force of the earth's gravity.[52]

The most interesting thing about plankton, however, is not their speed, but the food they transform in order to make it available to the rest of the marine ecosystem. Large animals do not feed directly on photosynthetic plankton such as microalgae or diatoms. In the ocean, unlike on land, there are very few grazers. Marine predators therefore need organisms to act as a bridge between the world of plankton and the animal world. Our copepods act as the crucial 'go-between': they convert the biomass produced by photosynthetic plankton into food available for the higher levels of organisms. They are therefore essential conveyor belts in the food chain.

The types of copepods present in an area therefore determine the animals that are found there. For example, large-scale population movements of cod in the North Sea are due to migrations of copepods in search of diatoms that only develop in colder waters. The disappearance of cod from these regions – where they were present for centuries – has led to tragic ecological and social upheaval.

Likewise, when copepods suffer from nutrient deficiencies,

these are passed on to their predators. Conversely, a balanced marine environment promotes the proper distribution of nutrients and the overall health of the entire ecosystem. This is why nutritionists recommend eating fish rich in omega-3: these essential fatty acids are produced by diatoms, but it's copepods that make them available, by incorporating them into the food chain. Basically, a fish is nothing but packaging for the end consumer.

A huge quantity of diatoms and other microalgae are ingested every day by copepods, thus limiting their proliferation and preventing them from becoming invasive. However, both copepods and diatoms are fragile organisms. Favourable conditions promote rapid reproduction, whereas unfavourable environments may cause their disappearance, with terrible ecological consequences.

Copepods do more than just make the resources produced by photosynthesis available. They also take part in the biggest daily migration on our planet.[53] Most of them come to the surface at night to feed, under cover of darkness to avoid being detected by the marine animals that eat them. When daylight comes, they dive down to the depths to escape the light and predators. These mass migrations are vital for the mixing and oxygenation of our ocean.

They also have a significant effect on carbon sequestration.[54] By feeding at the surface and defecating in the depths, these miniscule crustaceans transport and sequester carbon towards the seabed.[55] This process is essential for limiting global warming.

Copepods, like diatoms, are excellent indicators of water quality; their mass mortality is often the first warning sign of the deterioration of an ecosystem. But it is possible to anticipate these crises: because copepods are transparent, it is possible to monitor the state of their digestive system. A well-filled intestine indicates that the individual is in good health. If it is empty, then the marine environment is under threat.

Cambrian Explosion: Biology's Big Bang!

Between 541 and 530 million years ago, life on earth experienced a spectacular acceleration: the 'Cambrian Explosion'. The event lasted only 11 million years and marked a turning point which eclipsed the previous 3.8 billion years of the history of living things. The Cambrian Explosion gave rise to radically new forms of life: the first legs, gills, claws, eyes and skeletons.

This evolutionary leap was the result of an immense planktonic diversity promoting the circulation of nutrients, the presence of oxygen and favourable temperatures. Photosynthesis played a key role in producing the necessary material, while competition between predators and prey, or hosts and parasites, stimulated rapid coevolution.

Before the Cambrian Explosion, the first non-plant multicellular life forms were dominated by the Ediacaran fauna, most of which resembled huge air mattresses. After inhabiting the seabed for 80 million years, they mysteriously disappeared. Only jellyfish survived. During this explosion, more complex forms of life appeared: sponges, anemones and other arthropods. The latter, the ancestors of insects, crustaceans and spiders, played a key role in this diversification. They were able to develop exoskeletons to protect themselves and were the precursors of fish, reptiles, birds, mammals and... humans!

4.5 billion years ago

4.5 billion years ago	Formation of the earth
4.2 billion years ago	Appearance of life on earth
2.5 billion years ago	The Great Oxidation Event Oxygen eradicates virtually all life on earth
600 million years ago	Appearance of multicellular life "Biological Big Bang" Life is marine only
540 million years ago	Cambrian explosion Life emerges from the ocean Evolution of life onto emerged lands

Now

This rapid and unprecedented development was made possible by a combination of unlikely conditions: the earth's position at an ideal distance from the sun, the shape of its orbit, the presence of water, an atmosphere that filters UV rays, and the creation of oxygen, ozone and a lithosphere. The moon, by stabilizing the earth's axis of rotation, has also limited extreme climatic variations. And above all, there was the constant good fortune of never being in the path of one of the billions of meteorites speeding across the universe.

On the face of it, the likelihood of all these factors coming together over time is remote. But it's not all down to chance. While living things have managed to adapt, they also managed to create a large number of these conditions (by producing oxygen and ozone, for example), in order to ensure the conditions for their evolution. The Cambrian Explosion was simply the culmination of 4 billion years of meticulous preparation.

Life on earth had one final stage to overcome. The territories that were still uncharted because they were too hostile for our adventurous plankton: the continents, and dry land.

Krill on the menu for whales

They can sometimes be seen from the sky, looking like pink clouds drifting between white glaciers. Some krill swarms bring together several billion individuals to form a mass of several million tonnes spread over 450 km², and 100 metres thick. That's a density of up to 20,000 individual krill per cubic metre.

Despite its modest size, the krill has a genome up to eight times larger than our own. In Norwegian, *krill* means 'small fish fry' and includes more than 80 species of crustaceans that are just a few centimetres in size. They are the favourite food of the giants of the ocean: blue whales devour up to four tonnes of krill a day. Penguins, seals, squid, bears, albatrosses and many birds also feed on it.

Euphausia superba, or Antarctic krill, is the largest (6 centimetres) and most widespread species of krill. This alone explains the migration of whales towards the polar zones. With 600,000 billion individuals concentrated mainly around Antarctica and a global weight of 500 million tonnes (five times the volume of the world's fisheries), krill represents the largest animal population on earth, but also the largest multicellular biomass on the planet.[56]

Like copepods, the various species of krill also act as go-betweens, grazing on primary matter-producing plankton and feeding larger marine animals, both freshwater and saltwater. Hence their nickname, 'cattle of the sea'. However, krill differ from copepods in terms of size (they are far larger) and also in terms of their life cycle: they can live for several years, whereas a copepod doesn't live longer than a few months. Plus, krill moult around every 15 days. This shedding of their old shells contributes to sequestering carbon in the deep sea.

Another key innovation of krill is their second eye, which makes them excellent hunters. A remarkable attribute, especially since, during periods of scarcity, their body shrinks while their eye remains the same size. So if krill have large eyes, this indicates that the species is suffering and lacking food. Their survival, and that of all the species that feed on them, is therefore threatened. Despite there being an abundance of krill, they are threatened by overfishing.[57] Rich in protein, omega-3 and other nutrients, krill is therefore a product of choice for aquaculture and oil production, consumed in northern countries and renowned for its therapeutic and prophylactic properties. After having long been limited by technical and biological constraints (the animal tends to get crushed, forming a mush in the nets and releasing a toxin that makes it unfit for consumption), krill fishing methods have 'progressed' to the point that today, huge trawlers catch up to 300 tonnes of krill per day in Antarctica. The result is a drastic reduction in the food available for large marine animals. To date, the Norwegians have taken the lion's share, fishing 70 per cent of the world's krill, and the Norwegian company Aker BioMarine now controls 60 per cent of the world's krill oil production.

Despite numerous alarming reports by NGOs and the

IPCC, Europe's five largest salmon producers are still pur-
chasing krill meal on a massive scale, and the establish-
ment of marine protected areas at the poles is coming up
against economic and geopolitical obstacles, particularly
from Russia. The FAO itself recently acknowledged '[that
the scientific basis of] management of the krill fishery is
poor and that additional information on the behaviour of
this species and fishing statistics are absolutely necessary.'[58]

Krill is also threatened by ocean acidification. The
increased carbon content of the atmosphere, and therefore of the water, alters pH and increases acidity, eventually
attacking and even dissolving the shell of the krill.

Finally, polar ice melt is causing a decline in krill and
a proliferation of 'salps' (distant cousins of jellyfish,
although taxonomically closer to humans), that devour all
the resources in the area, hindering the renewal of krill. The
disadvantage is that, unlike krill, these gelatinous organisms
do not feed the upper links in the food chain. If krill and
copepods disappear, a huge part of the marine ecosystem
could also become extinct.

Plankton also invented eyes!

The eye is undoubtedly one of the most extraordinary inventions of life, the result of four billion years of evolution. It originated in simple cells that were drawn towards light. After billions of failed attempts over several million generations, the feat was achieved: grouped around a brain, these cells ended up forming eyes.

The eye remains the ultimate challenge when creating prosthetics or designing cyborgs. We can send our fellow creatures into space, but are still unable to artificially recreate these ocular spheres. Darwin was one of the first to question such complexity, as these globes ran counter to his theory on the subject. He humbly acknowledged this in 1859, and his detractors used the complexity of this spherical organ to refute his hypotheses.

How could an aggregate of cells form these two spheres capable of capturing and filtering light through the iris, the cornea and the pupil, of correcting its focus and of transforming photons into electrical signals that can be interpreted by the brain? A seemingly impossible process. And yet one that has been achieved through evolution. The history of the eye is difficult to trace because these soft masses are poorly preserved in fossils. However, by observing the food chain of plankton, researchers have gleaned some precious clues. Unlike life on land, underwater life is like an archaeological archive, revealing the order of the stages of this evolution. To understand how the need to invent eyes arose, we must go back to the creator of everything around us: the sun.

It all began with primitive organs: phytochromes, light sensors present in certain diatoms which enable them to optimize photosynthesis and find their way through rough water. These ancestors of our eyes are also used by plants to direct their leaves towards the sun.[59] It has long been observed in plants that these sensors only perceive red light, but this is absorbed by water and is therefore absent from the ocean.[60] Only blue light is reflected. Hence the colour of the ocean when plankton are not proliferating at the surface. But then, why would our diatoms have developed such a sensor?

Thanks to data collected by the *Tara* expedition, researchers discovered that these sensors in diatoms were much more efficient than those in plants and captured all light, including red. These more advanced phytochromes have enabled diatoms to orient themselves according to variations in light and to modulate their depth in the water so as to optimize photosynthesis.

For a long time, people also wondered why these phytochromes were only present in diatoms living outside of the tropics. Equatorial waters are relatively stable, while the waters in temperate zones are more agitated. Diatoms use these light sensors to get their bearings in moving environments. So this is how photosynthetic plankton invented a very primitive form of vision.

Subsequently, evolution forced their consumers to find ways to move towards these prey, and therefore towards the light.

Copepods were among the first animals at the bottom of the food chain to develop a black spot in the middle of their foreheads. This 'naupliar eye' allows them to perceive light. A little higher up the chain, the microcrustacean *Artemia Salina* develops two more eyes, but only in adulthood.

Krill have two eyes from birth, suggesting their 'hunting' nature. An organ that can tell the difference between day and night, or distinguish between what might be a passing shadow, predator or prey, provides a substantial selective advantage. Eyes rapidly became more complex. Between predators and prey, it's a matter of who evolves the fastest, each advancing their new asset, forcing the other to evolve in turn. In this dizzying game of chess, the survival of the species is essentially at stake. Crustaceans and fish came to form highly evolved eyes, similar to those of mammals. Over the millions of years that followed, the sophistication of eyes continued to diversify within the animal kingdom – there are no eyes, as such, in plants or fungi.

In insects and crustaceans, compound eyes appeared, made up of multiple optical elements, each capturing a part of the image, thus allowing a wide field of vision. Spiders and molluscs have ocelli, simple eyes that detect light without forming detailed images. Octopuses and squid have multi-lens eyes that allow them to see in a variety of light conditions and with great precision.[61] Snakes have developed tubular eyes, i.e. they are elongated and perfectly adapted to scanning objects over a long distance despite having a reduced field of vision. Finally, owls and cats have evolved eyes that can see in the dark, which are incredibly useful for these nocturnal predators.

The eye continues to evolve, even within our own species. 8,000 years ago, a *Homo sapiens* from the shores of the Black Sea underwent a mutation, altering the OCA2 gene on chromosome 15. This unusual feature gave him blue irises. We don't know who the individual was, but they were certainly unique. Everyone with blue eyes has a trace of them in their genetic code. Scientists are still wondering why this recessive evolutionary trait was passed on, as it offers no advantages and is not even well suited to prolonged exposure to the sun. However, in Northern Europe, where the majority of Caucasian descendants migrated, over 80 per cent of the population now has blue eyes. This is a counter-intuitive anomaly for a recessive gene which would normally be expected to disappear over the course of time with genetic mixing.

Researchers eventually concluded that these blue eyes are simply an 'evolutionary advantage for reproduction'. The marked aesthetic preference for this azure-eyed family has enabled this mutation to make its way to us. Some of its members will certainly appreciate this theory.

A commando unit of plankton attacks dry land

600 million years ago, a daring group of planktonic creatures achieved a feat: they colonized dry land. Until then, life on earth was limited to the ocean as the emerged parts were too hostile and arid, saturated with sun and oxygen. Underwater, the light was filtered and the temperature remained stable. To prepare for landing on such an inhospitable environment, plankton selected an elite unit: a symbiotic alliance between microalgae and fungi rose to the challenge. The success of the mission required the help of certain bacteria, protists and other planktonic viruses. Carried by winds and tides in droplets of water, these pioneers parachuted onto rocks and formed an exceptionally resistant symbiosis: lichens.

Through photosynthesis, algae capture carbon and produce organic matter that feeds fungi.[62] In return, the fungi cover the algae with moist protective filaments and supply them with minerals extracted from the rocks. This symbiosis has created incredibly resilient organisms. 500 million years later, lichens continue to spread to every corner of our planet.

These surprising organisms are capable of surviving in extreme conditions. From the poles, to the mountains, to the coastlines, they can withstand temperatures ranging from 90°C to -200°C.[63] One experiment has shown that they can even survive for weeks in the vacuum of space.[64] Today, 20,000 species of lichen inhabit our planet. They enabled the soil to become vegetated, creating fertile ground for the arrival of other forms of life. After the infantry, the heavy artillery landed on earth around 450 million years ago: green marine macroalgae, also transported by the winds and tides, took root in these new fertile territories, creating terrestrial plants, forests and meadows.[65] Then came the cavalry: the first animals, notably arthropods, which traded their fins and gills for legs and lungs. This mutation can be seen in certain so-called 'pulmonate' fish, which still have both gills *and* lungs.

This terrestrial evolution would not have been possible without the oxygen emitted by photosynthetic plankton. Oxygen is converted into ozone when exposed to UV light, protecting terrestrial organisms from direct exposure to solar radiation. Nevertheless, some species did not survive these major mutations, while others beat a retreat. And so the ancestors of whales, dolphins and dugongs, after trying for tens of millions of years to adapt to dry land, finally returned to live in the ocean.26 Perhaps one day, to escape temperatures when they become unbearable, we will also return to the ocean in a slightly different form...

Paradoxically, this sudden expansion of the plant world, which enabled life to develop on earth, also upset the planetary balance and led to the destruction of part of its biodiversity. The first major mass extinction of the multicellular world, known as the Ordovician–Silurian extinction events, was a consequence of the oxygen produced by terrestrial vegetation. The large-scale sequestration of CO_2 by plants led to a gigantic glaciation that destroyed 85 per cent of marine life. Yet, with every mass extinction, life eventually triumphs and takes a new direction. The history of life is characterised by these alternating cycles of evolution and extinction. Most species disappeared, but those that survived adapted to evolve towards ever greater complexity. In doing so, they also changed their environment, often to the detriment of species already present. Through these transformations, evolution gave birth to increasingly complex life forms.

These mutations also created a particularly efficient primate: humans. Although we may have forgotten, we come from the planktonic world. Let's remind ourselves: we begin life in the form of a microscopic cell equipped with flagella, like the majority of plankton. We float in a protective liquid for the first nine months of our lives, just like our ancestors drifting in the ocean. Moreover, from birth, we create symbioses with microbes, fungi, protists, bacteria, viruses and archaea, all descendants of plankton which, from inside our intestines, protect us in exchange for digestive services. As with lichen, this relationship allowed us to leave the aquatic world and conquer the terrestrial world. And once again, as primates we've had a considerable influence on the history of life. But there is one major difference in our era: for the first time in the history of life, we are fully aware of the mass extinction that is happening.

From the *Alien* monster to the world's longest animal

Zooplankton include a fascinating diversity of organisms, from temporary larvae to single-celled protozoa. These larvae go through complex metamorphoses and, depending on their age, feed on various plankton in such sophisticated patterns that it is complicated to grow them. Our inability to breed tuna in captivity stems from our inability to understand and feed these larvae.[66] That's why it's much easier to just catch juvenile tuna that have emerged from the larval stage and then fatten them up.

Protozoa are among the most basic zooplankton on the evolutionary scale, and include a large number of very different species. Being single-celled, they generally feed on bacteria, microorganisms or organic debris. While the vast majority of these are harmless, some are known to carry pathogenic species causing diseases such as malaria or dysentery, while others make up the group of amoebae.

Among planktonic crustaceans, the *phronima* genus has left its mark on the history of sci-fi, firstly inspiring the French cartoonist Jean 'Moebius' Giraud's surreal characters, and then serving as the model for the monster in the film *Alien*. This small see-through predator, measuring only a few dozen millimetres in size, has a large head with bulging red eyes and sharp pincers. It can make itself almost invisible to its predators and parasitizes salps and sometimes jellyfish, laying its eggs inside them. The parasitoid and its offspring then slowly devour their host from the inside, while remaining safe from their usual predators.

Other surprising creatures inhabit the ocean, such as *Clione limacina*, a small aquatic slug that is said to have inspired Pokémon.[67] But one of the most unusual is the siphonophore![68] This animal resembles a long gelatinous snake that can measure up to 120 metres. This plankton usually moves its tentacles silently through the depths, its soft body allowing it to withstand the enormous pressure down there. Its structure is fascinating, as it is a colony comprising several thousand individuals. These perfectly identical 'zoids' have very different functions and morphologies: locomotion, reproduction, digestion, defence and predation. They are all connected by a sort of umbilical cord that acts as a backbone, the whole forming a mega-organism that is unique in the world; a floating city of sorts. And beware, anyone who comes across it, as its tentacles are armed with terrible stinging cells.[69]

One of its most feared representatives is *Physalia physalis*, or the Portuguese man o' war. Originating in tropical seas, it is increasingly found on our coasts, much to the despair of fishermen as its grip can be painful, even deadly. In 2024, beaches in Catalonia were closed when enormous numbers of these individuals showed up. In French, its name is the equivalent of 'sea bladder', because of the large float, or pneumatophore, that protrudes from the surface of the water and enables the wind to transport it over long distances. It has proliferated recently on European coastlines, much like jellyfish have, as both thrive in an ocean disturbed by human activities. An invasive cousin that could turn out to be one of the biggest winners in the climate crisis.

Planet of the jellyfish: should we fear them or learn from them?

Jellyfish are perhaps best known for the painful burns caused by the venom they emit to immobilize their prey before devouring it. In humans, these toxins are usually relatively harmless, but some are fatal, like the 'box jellyfish' that lives in Australia and South-East Asia.[70] This little jellyfish, 20 to 30 centimetres in diameter and virtually invisible in water, contains enough poison to kill sixty people.

For some years now, various species have been appearing in the form of 'jellyfish soups' along beaches, with clusters of up to 800 individuals per cubic metre of water. In 2007, an Irish salmon farm was assailed by a gigantic shoal of billions of jellyfish, 13 metres deep and covering an area of 26 km². For nearly seven hours, the gelatinous creatures methodically poisoned more than 100,000 salmon, devouring their carcasses while the boats were immobilized in this viscous soup and the fish farmers looked helplessly on.

The year before, in Japan, jellyfish weighing 200 kg and measuring 2.5 metres in diameter arrived from the China Sea. They invaded the waters, overturning boats and tearing nets, and forcing fishermen to suspend their activities for several weeks. These stories illustrate just how widespread the outbreaks of these survivors have become. They have withstood every cataclysm in the history of life, and look set to continue to do so.

Although we often assimilate the flaccid, translucent forms floating in the ocean to jellyfish, gelatinous organisms actually include very different animals belonging to distinct branches of life. In other words, at the genetic level, there is as much similarity between jellyfish and salps as there is between humans and insects. The word 'jellyfish' is actually a misnomer which corresponds to a stage of development of certain cnidaria. They initially exist as polyps, encysted in the sediment, before detaching and becoming planktonic in their 'jellyfish' stage. Calling such a wide range of organisms 'jellyfish' is like using the term 'adult' for every living thing on earth. But the term 'jellyfish' is now so widespread that, for the sake of simplicity, we will carry on using it in this book.

Almost three quarters of the 1,500 known species are microscopic, while the largest can reach up to 2 metres in diameter and an overall span of 36 metres including tentacles. But what all these creatures belonging to the plankton family have in common is that they drift with the current over very long distances.

While most animal species are disappearing as humans destroy the planet, jellyfish seem ready to survive another mass extinction. After all, they've already made it through the first five! Far from being negatively impacted by the amount of waste, nitrates and other agricultural fertilizers we dump into the oceans, jellyfish actually benefit from them. Excess nutrients stimulate the proliferation of photosynthetic plankton, and therefore the zooplankton they love to eat.

Similarly, the CO_2 emissions that are warming the oceans provide them with a favourable environment for development. The resulting acidification of the oceans also prevents many crustaceans and molluscs from forming their shells, making them easier for jellyfish to digest. Even the billions of plastic bags floating in the ocean work out well for them: large predators such as turtles and seabirds ingest them, mistaking them for jellyfish, and end up suffocated by our chemical packaging. Free of these troublesome hunters, jellyfish can also take advantage of the buoyancy of our microplastics to deposit their polyps (larvae) and conquer new territories.

Finally, overfishing and large industrial trawlers are freeing jellyfish from both their predators (tuna, mackerel, sunfish, sharks, turtles) and their competitors (the crustaceans and other marine fish that consume the same copepods as jellyfish). This is a double blow for our fish and crustaceans: not only does overfishing decimate the predators of jellyfish, but in addition, the booming population of jellyfish also feed on their descendants drifting in the ocean in the form of planktonic larvae. This phenomenon is accelerating the imbalance in the marine ecosystem, leading to what some experts – including Fabien Lombard of the CNRS – are already calling an 'ocean of jellyfish'.[71]

Currently, in Namibian waters – once exceptionally rich in marine animal life – the biomass of jellyfish is estimated to be two and a half times greater than that of fish. The main cause of this drama currently unfolding has been traced back to the overfishing of sardines. In south-east Europe, one variety of ctenophores (comb jellies) multiplied and devoured all the living resources of the Black Sea, turning it into a gelatinous broth. The Baltic Sea suffered a similar fate. Proliferations of the same Ctenophora can be found in the Gironde estuary in France, while those from the Black Sea travel as far as Barcelona, on the other side of the Mediterranean.

They are not affected by eutrophication, lack of oxygen or even lack of food. Jellyfish rise to all challenges, and are even capable of going dormant for long periods in order to survive. *Pelagia noctiluca*, a jellyfish known as the mauve stinger, proliferates in the Mediterranean and can go for a month without feeding. If necessary, they can even become cannibals and devour each other.

They can also sometimes be quite inventive. In the Palau Islands, east of the Philippines, a 'jellyfish lake' has formed: two species became trapped there following the obstruction of a tunnel leading to the ocean. In the absence of any predators, the jellyfish multiplied so well there, that there are now over 10 million of them. When all the food in the lake had been consumed, they formed a symbiotic association with a microalga. These jellyfish transport their microalgae to the sun so that they can take advantage of the light and produce organic matter that will serve as food for their gelatinous partners.

But jellyfish are not solely harmful; they actually play an essential role in the balance of marine ecosystems. By feeding on small fish and zooplankton, they regulate aquatic populations. Eradicating them is therefore not the answer. Jellyfish, which first appeared 650 million years ago, are also champions at adapting to climate change.

These indestructible creatures have survived extinctions that have devastated up to 90 per cent of life on earth: ice ages, planetary warming, predators, lava flows, volcanic ash fumes and comets that wiped out the dinosaurs. Nothing could stop them!

Surprisingly, these creatures are extremely fragile, and they are killed by the slightest impact. This is why so many dead jellyfish are often found washed up on beaches. Collectively though, they are incredibly resilient because they are capable of optimizing all their metabolic functions to the max! Devoid of heart, brain, blood and lungs, they are 98 per cent water, and everything they do is focused on a single objective: the survival of the species.[72]

Their umbrella-shaped bodies, made up of two thin layers of cells separated by an aqueous gel, can store the little oxygen they need. This allows them to diffuse oxygen and carbon directly through their epidermis. Although they cannot swim against the current, their tentacles allow them to move easily through the water column to feed. Their only internal organs are reproductive gonads. This is essential given that they are capable of reproducing abundantly, with an individual producing up to 19,000 eggs per day.

When swarms of jellyfish show up, it's a real disaster for aquaculture. Waste from farmed fish makes aquaculture areas rich in nutrients, while their infrastructures provide the ideal surfaces for polyps to attach to. A few years ago, in Taiwan, in order to combat this scourge, the government closed down all the fish farms for several months, which brought the jellyfish outbreak to a sudden halt.

This scourge is a global one, and in a few years the entire

economy built around fishing, aquaculture and seaside activities, and even energy production, could suffer as a result. Because beyond their impact on biodiversity, jellyfish also disrupt human activities, for example by clogging the cooling circuits of nuclear power plants and thus jamming the reactors, as seen in recent incidents in South Korea, the United States and Sweden.

Some of them even go so far as to be immortal. The now famous *Turritopsis dohrnii,* also known as the immortal jellyfish, has invaded all of the world's waters. When it is suffers stress – whether it is amputated, injured, sick or close to death – this species returns to its larval polyp state and regenerates itself, like something out of a sci-fi film. This cycle can be repeated indefinitely, making this jellyfish biologically eternal. It would be like if we, after a century of existence, returned to the embryonic state in our mother's womb to start a new life over again. Very recently, it has also been discovered that some gelatinous animals are capable of fusing two injured individuals together to re-form a single healthy individual.[73] The new organism then has the DNA of the two original individuals. This case appears to be unique in the animal world.

Needless to say, some researchers have tried to transpose these processes to humans. Shin Kobuta, a marine biologist specializing in this 'immortal jellyfish', believes that 'the human mind is not ready for eternity'! Eternity is long, 'especially towards the end', so it is likely that the absence of a brain is a definite evolutionary advantage for the jellyfish, as it prevents it from thinking too much about its metaphysical condition.

Like global warming, some people claim that these jel-lyfish blooms correspond to cyclical phases of our planet. However, the acceleration of the phenomenon and the significant connections to human activities are increasingly challenging this kind of reasoning. Others are seeking to eradicate jellyfish populations, for example in South Korea, where they have manufactured machines that can crush 900 kg of jellyfish per hour.[74]

Instead of exterminating them and potentially creating ecological imbalances, why don't we try to find sources of inspiration in their extraordinary biological make-up?

Even if we're unable to prevent the extinction of our spe-cies and that of most of the living world, we can take com-fort in remembering that jellyfish will most likely survive. And if they continue to rapidly evolve, adapt and reproduce, perhaps they will eventually dominate the planet. In fact, they did so for several tens of millions of years, just before the Cambrian explosion. After all, at the time, who could have guessed – watching the first vertebrates, unimpressive little fish creeping along the ocean floor, feeding on sedi-mentary debris – that this would eventually lead to majestic whales and to human beings walking on the moon, with incredibly powerful super-computers at their fingertips? So perhaps the future could be jellyfish?

The plankton paradox

In ecology, the competitive exclusion principle, also known as Gause's law, states that under natural conditions there cannot be more consumers than the available resources to feed them in a given environment. Yet plankton defies this logic! As early as 1961, the father of modern ecology, George Hutchinson, observed that certain areas of the ocean have a plankton density far greater than the resources available should allow. And so, the 'plankton paradox' came about.

The complexity of the food chain makes this phenomenon difficult to explain, but one thing is for sure: everything depends on photosynthetic plankton which, nourished by sunlight, CO_2 and a few nutrients, thrive and feed the rest of the trophic pyramid.

Unlike terrestrial biomass, which is dominated by plants, oceanic biomass represents only 1 per cent of global living organisms, yet produces 50 per cent of the planet's oxygen. So 1 per cent of all living organisms produce as much oxygen as the other 99 per cent. This observation is even more impressive given that only a fifth of this ocean biomass produces oxygen (fish, crustaceans and shellfish do not). So these microorganisms, which account for 0.2 per cent of the earth's biomass, or a gigatonne, produce as much oxygen as the 460 gigatonnes of plants, which account for 83 per cent of the earth's biomass.

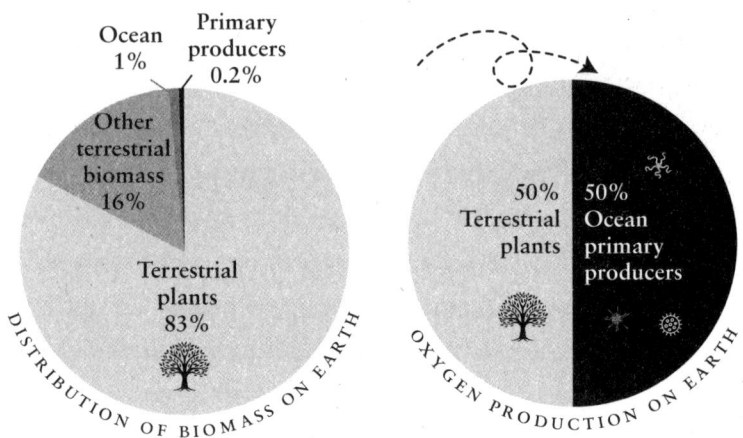

Marine producers provide half of all oxygen

This efficiency depends on two fundamental characteristics: firstly, the lack of organic matter storage because, unlike land plants, plankton have no stems, roots or branches. Each cell is entirely dedicated to its photosynthesis and no reserves are created. Secondly, the productivity and renewal rate of photosynthetic plankton are ultrafast. Whereas on land, a forest regenerates over several decades, in the ocean, all of the photosynthetic plankton can be renewed in five days. To illustrate this productivity, the US research centre Bigelow calculated that if you were to align all of the microalgae in the world's oceans, you would end up with a 7 cm x 30 cm plank stretching 386,000 km, from the earth to the moon.[75] These two specificities logically lead to the pyramids of life being very different for the land and ocean. On land, a cow consumes a quantity of grass that far exceeds its own weight. Its metabolism uses part of that to survive and creates reserves with the rest. The cow will offer these reserves to the upper level, forming a traditional ecological pyramid with a wide base, and biomass that decreases at each level above that.

In the ocean, the situation is very different. With the exception of macroalgae and aquatic plants – whose mass is relatively negligible – most photosynthetic organisms that produce organic matter do not store reserves. They are born, frantically reproduce and are then consumed. This is not the case for their consumers, the protists and multicellular animals of the higher levels that build up their metabolism by devouring them. And each new level of the pyramid stores more organic matter than the one below. So, at a given moment, we find ourselves with a situation that is very counter-intuitive for us.

This particular dynamic – still poorly understood – could change our ecological and climate models. Many oceanography experts believe that IPCC studies still ignore too many of these specificities in the ocean, leading to significant errors in climate projections. Understanding the pyramid of life, based on plankton, is essential if we are to anticipate the future of our ecosystems.

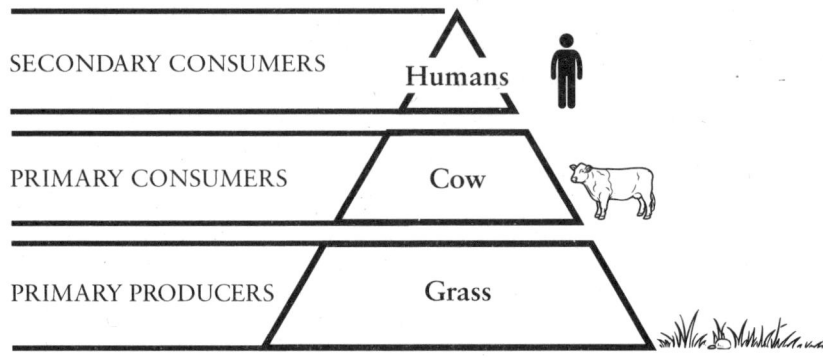

Terrestrial trophic pyramid

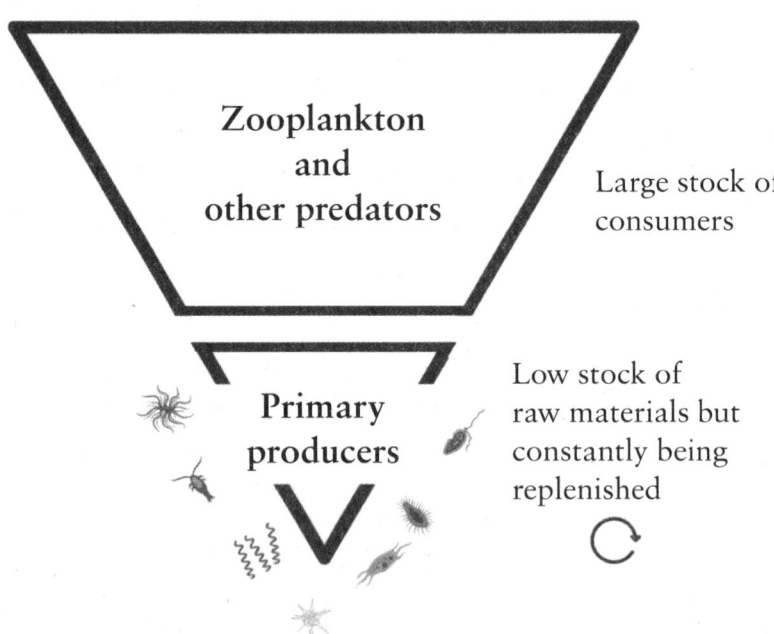

The inverted structure of the trophic pyramid in the ocean

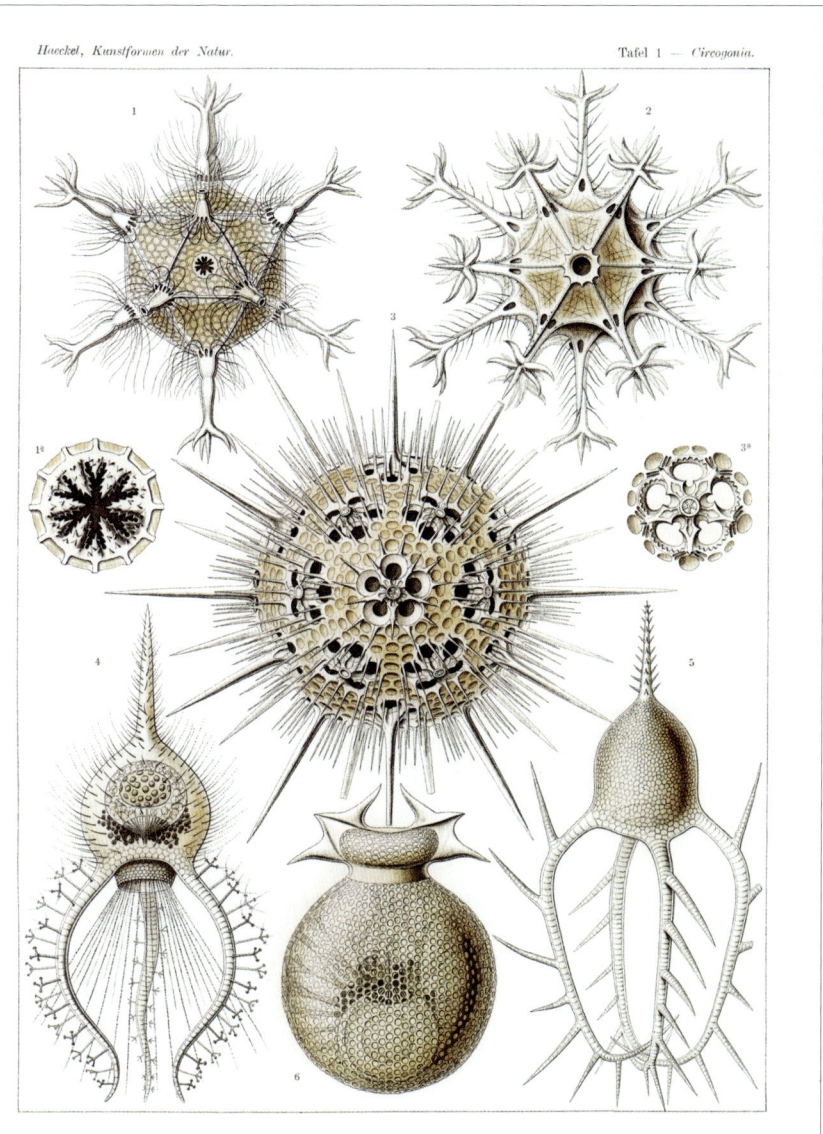

Phaeodaria. — Roßrstraßlinge.

An example of Ernst Haeckel's intricate illustrations
of radiolaria, from *Art Forms in Nature* (1904).

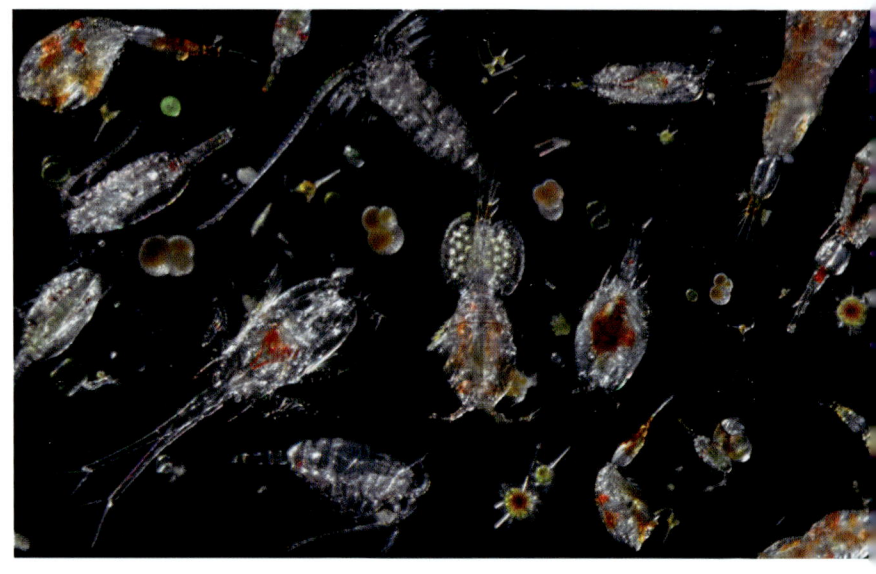

Copepods and protists
(photo: Christian Sardet).

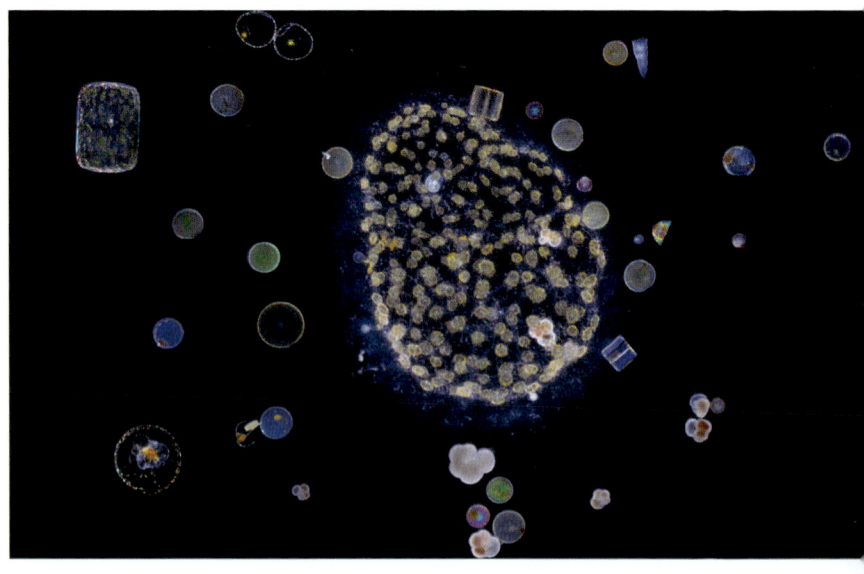

Unicellular and colonial protists
(photo: Christian Sardet).

The author with a plankton net aboard the schooner *Tara*,
dedicated to exploring the invisible life of the oceans.

Jellyfish blooms – dense aggregations of jellyfish
formed under certain environmental conditions (*Shutterstock*).

Open-air spirulina cultivation ponds.

Plankton observed under the microscope.

Microalgae cultures.

The author en route to the Arctic Circle, where plankton ecosystems are rapidly changing under the effects of global warming.

Giant siphonophore.

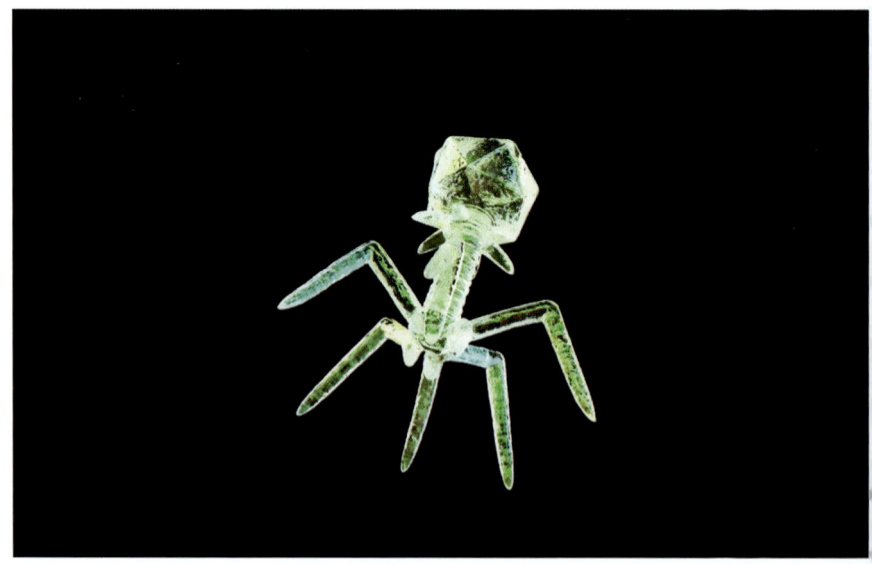

Bacteriophage virus
(*Shutterstock*).

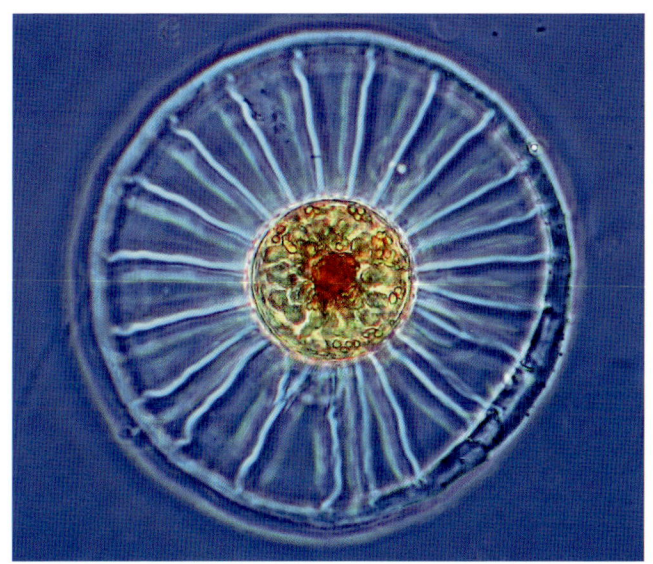

Wagon wheel diatom.

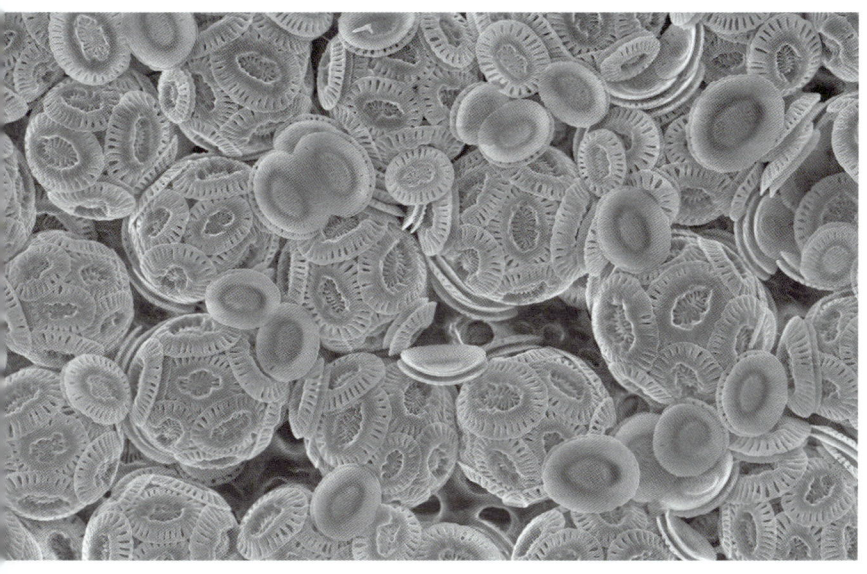

Coccolithophores, microscopic algae whose remains
have formed vast chalk deposits over geological time.

Bioluminescent plankton covering the sea
(photo: Stephan Sprinz/Wikimedia Commons)

A toxic algal bloom, or 'red tide'
(photo: Kai Schumann//Wikimedia Commons).

Early 21st century: *Tara* and the first great expeditions into the invisible world

It all begins like a novel: with a pirate attack. On 6th December 2001, Peter Blake, a sailor from New Zealand and winner of the America's Cup, was on a mission for the UN on the Amazon river, when he was shot in the back by assailants and killed.

His beautiful schooner, designed by Jean-Louis Étienne, remained docked for two years before it was bought by agnès b. in 2003. The fashion designer founded the eponymous clothing brand which, with nearly 300 million euros in turnover and more than 2,000 employees worldwide, has become one of the flagships of French fashion. The woman who was known for inventing the iconic 'snap cardigan' comes from a family that is passionate about the maritime world. Agnès and her son Étienne acquired the boat to continue Peter Blake's fight. But in a slightly different way. She renamed the schooner *Tara* and went on to become one of the most enthusiastic supporters of ocean research, and in particular plankton research.[76] Designed to withstand ice, in 2006 *Tara* set sail for the Arctic: a scientific expedition was invited onboard to study the effects of climate change. The initiative attracted the attention of Éric Karsenti, a renowned biologist whose dream was to reproduce Darwin's legendary nineteenth-century exploration aboard the *Beagle*. But this time, the observations would solely focus on plankton. In 2009, he convinced the *Tara* team to launch a unique project, the *Tara* Ocean Mission.

After months of preparation and technological innovations (notably the rosette, a device for collecting water samples in deep water)[77, 78] – which would soon become universal landmarks in the field of plankton research – the expedition set off, for several years. *Tara* is the first large-scale project to study plankton *in vivo*. Oceanographers, biologists, climatologists, marine palaeontologists, chemists, artists and journalists, with the support of the CNRS and international institutions, take turns on board to offer a multidisciplinary approach.

The entire team works in collaboration with a network of laboratories located around the world that collect the samples. The logistics are complex but they work perfectly. 200 researchers work on the schooner to create the world's largest open-access database of information on plankton. Many researchers have joined the adventure, as part of the '*Tara* generation'.

Thanks to technological advances in DNA sequencing, the expedition has taken on an unprecedented scale. What seemed like a utopian dream has become the largest genetic sequencing project ever carried out on marine organisms. Over two and a half years, *Tara* travelled 125,000 kilometres, collecting over 40,000 samples, identifying more than 200,000 marine viruses, 100,000 microalgae, 500,000 bacteria and more than 150 million new genes. By way of comparison, our human body is made up of just 20,000 genes.

In addition to this genetic database, *Tara* has improved the ways in which we can measure the impact of human activities on communities, and the interactions between communities. This avenue was further explored by another mission, '*Tara* microbiome', conducted between 2020 and 2022 around South America and Africa. The scientific impact is huge: more than 1,000 publications, including some in *Nature*, *Cell* and *Science*. This initiative has also inspired others. In 2010, the *Malaspina* expedition in Spain explored the deep sea. Other expeditions continue to draw inspiration from the *Tara* concept, combining scientific research, awareness-raising and ocean preservation to study marine ecosystems, biodiversity and the impacts of climate change on the ocean.

Today, with its 'Polar Station' mission, *Tara* is focusing on polar regions. In parallel, *Tara* is raising awareness among millions of people, particularly the younger generations, about the ecological challenges of the ocean environment, especially those related to plankton.

The missions are ongoing, and studies on plankton continue to progress, but there is still a colossal amount of work to be done on fundamental research. While *Tara*'s database allows us to identify some of the populations and link them to a place or environment, we still need to learn about and understand this underwater biodiversity, its mechanisms, its behaviours, its interactions, its impact on the environment and climate. We are at a turning point in the history of life, where our species has never been so threatened with extinction, and where, paradoxically, humanity is beginning to glimpse the immensity of underwater and invisible life.

What if, at last, we could measure the depths of our ignorance – and, in doing so, save ourselves?

Plankton, satellites, robots, DNA and AI: heading for a major civilizational shift?

The nineteenth century saw major advances in botany thanks to the great explorations. These developments truly changed the world. But this research focused only on the visible world, excluding a whole invisible (microbial) world, and even more so the marine world. At the beginning of the twenty-first century, new technologies are opening up an unprecedented field of investigation into life on earth, heralding a new civilizational leap.

In the ocean, everything remains to be discovered. UNESCO estimates that only between 10 and 15 per cent of marine species are known (approximately 226,000 species out of a presumed total of around 1 million).[79] Currently, 2,000 marine species are scientifically described each year, and in the past decade more have been discovered than were known in total before. The study of plankton is even more recent and has accelerated more than any other field in the last twenty years.

It was not until the invention of the microscope in the seventeenth century that the incredible microscopic richness that nourishes underwater life was revealed. And it took until 1816 for an English naturalist to develop a net with a mesh fine enough to capture these tiny drifting organisms.

Darwin took his invention with him on the Beagle and was the first to understand the role of plankton in this food chain that is invisible to our eyes.

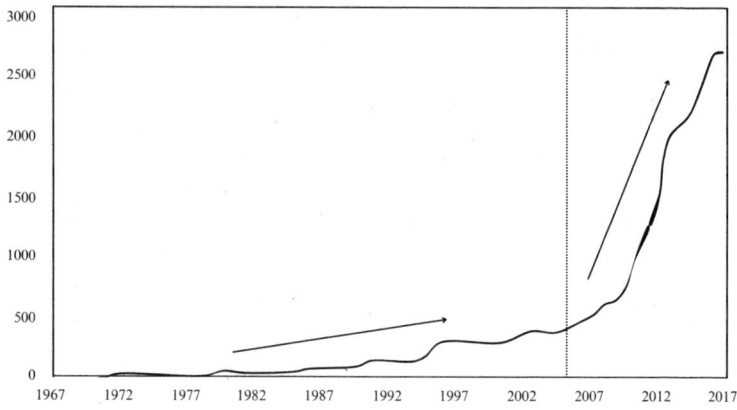

Number of scientific publications on plankton (1970-2017)
Source: Grima et al., 2013

Knowledge progressed further at the turn of the twentieth century under the influence of German researchers who were eager to challenge English naval supremacy. In 1887, the botanist Victor Hensen defined the bases of marine biology and invented the term 'plankton', from the Greek *planktos*.[80] His compatriot Haeckel, as we saw earlier, would follow in his footsteps.

In the interwar period, England launched the Continuous Plankton Recorder monitoring programme, which is still ongoing and measures the composition of plankton in all four corners of the globe. A small torpedo-shaped sensor is attached to boats to collect plankton. Every year, approximately 250,000 kilometres of ocean are sampled.

It is only in the last half-century, however, that new

technologies have really made it possible to start glimpsing the hidden part of the planktonic iceberg. For example, *Prochlorococcus*, the most widespread photosynthetic organism on earth, responsible for 5 per cent of global oxygen production, was discovered by scientists in 1986.

Satellite imagery now makes it possible to monitor the evolution of plankton by analysing the colours of the ocean. This spatial data is cross-referenced with offshore measuring equipment capable of detailing environmental conditions and filming microscopic plankton under the sea. Scientists also use underwater sensors to study the sounds emitted by plankton in order to recognize them, the way we do with birdsong.

In February 2024, NASA launched the PACE (*Plankton, Aerosol, Cloud, ocean Ecosystem*) program to observe the interaction between phytoplankton, atmospheric gases and clouds. The satellites usually work in interaction with robots floating in the ocean.

One example of this is the Argo floats, named after the ship Jason sailed in search of the Golden Fleece. These floats drift with the ocean currents and move vertically between the surface and the depths, measuring the temperature, the salinity of the water but also the biological conditions of the oceans, data which they transmit, at the surface, via satellites, to terrestrial reception stations. With over 4,000 floats worldwide, used by more than thirty different countries, the Argo program is now the most significant source of ocean data in human history.[81]

Subcellular microscopy allows us to study the processes that take place inside the cells of this drifting people.

Cryo-electron microscopy (cryo-EM), a technique combining rapid freezing and advanced microscopy, was awarded a Nobel Prize in Chemistry in 2017 and revolutionized the study of molecular structures.

DNA sequencing, combined with artificial intelligence, makes it possible to model the interactions between these thousands of species. The fact that we can combine satellite data with samples taken at sea and computational genomics is opening the door to a new understanding of life.

This approach is not just for oceans: freshwater in rivers, streams, lakes and even puddles is teeming with biodiversity and also offers incredible potential for biotechnological innovations.

The integration of this new potential and increased investment in research might allow us to better understand the entire living world, on land and at sea.

Because, given our limited knowledge of the ocean, it seems that life on our planet is a game whose rules we know practically nothing about.

Plankton and humans: an invisible massacre

Development of human activities
- Intensive agriculture and industrial waste
- Development of aquaculture and coastal infrastructures
- Emission of greenhouse gases

Socioeconomic consequences
- Decline in fish and aquaculture
- Economic losses for touristic and coastal activities
- Reduction in high-quality freshwater resources

Environmental changes
- Water warming
- Ocean acidification
- Ecological imbalances

Deterioration in water quality
- Nutrient enrichment (nitrogen, phosphorus)
- Hypoxia (lack of oxygen)
- Disruption of food chains due to the depletion of the plankton that feeds them

Proliferation of opportunistic organisms
- Proliferation of jellyfish and microalgae to the detriment of plankton supporting current ecosystems
- Obstruction of infrastructures (central, aquaculture)
- Threats to marine species

Cycle of interactions between human activities and ecological damage

While the current balance of planktonic communities has helped to shape the living world, it has also been largely modified by its evolution, and the most significant impact today is clearly caused by humans. We have violently exploited the planet's resources for our own benefit.

Firstly, climate change plays a key role in the ongoing upheavals within plankton populations. Rising temperatures and ocean acidification are altering the ability of certain plankton to survive and encouraging the proliferation of other species that are less compatible with our ecosystems. In addition, the increasing frequency of extreme weather events leads to floods laden with chemicals that disrupt marine biodiversity.

But climate is not the only factor involved. Everything that is dumped on land ends up in the water (oceans, rivers and lakes) and affects planktonic communities. In many developing countries, the lack of sewage treatment systems is leading to a massive influx of nutrients into coastal areas, threatening biodiversity. This phenomenon is also found in the most developed countries. In France, for example, as a result of poor wastewater management in the Finistère region, a norovirus contaminated oysters in 2022, causing digestive problems among consumers. Following the ban on the sale of all shellfish from the Bay of Morlaix, the regional council decided to only subsidize municipalities that had effective water treatment systems and to step up inspections. The crisis was quickly brought under control, oyster farming resumed and noroviruses disappeared, showing that solutions are therefore possible, even if these measures are costly.

Another major threat is our dependence on pesticides and

other agricultural inputs (herbicides, phosphates, nitrates). France alone consumes around 110,000 tonnes of pesticides a year, 99 per cent of which end up in rivers and then in the ocean, where they accumulate over a very long period of time.[82] Recent analyses have revealed traces off the coast of California of DTT, a pesticide that has been banned since the 1970s. The stakes are high because toxins accumulate throughout the food chain, reaching particularly high levels in seabirds and carnivorous fish (sea bass, salmon, sea bream, trout, etc.).

At the same time, agricultural inputs, such as phosphate and nitrate, promote the proliferation of nutrients, and therefore of invasive algae, thus reducing oxygen and light for other species in the ecosystem. Recent studies have clearly established the link between the development of intensive agriculture and the profound changes in plankton communities since the second half of the twentieth century. Once again, these disruptions benefit certain plankton that are undesirable or even harmful to human health, to the detriment of other plankton with which we have learned to live.

In Argentina, researchers have demonstrated the impact of glyphosate on planktonic communities in fifty-two lakes. In this case, the alteration of ecosystems is characterized by high water turbidity, an increase in cyanobacteria and a reduction in the diversity of microalgae. Glyphosate is just one of many examples. Reducing the use of pesticides and other agricultural products is a priority measure for ensuring the health of our land, our oceans and life in general.

The use of antibiotics in livestock farming also contributes to the development of antibiotic-resistant planktonic bacteria in water treatment plants. These bacteria then spread into nature and pass on their characteristics to insects and other animals.

Another source of disruption is ballast water related to shipping; this is the seawater used to fill the empty holds of ships to stabilize them after unloading cargo. When this untreated water is released, it encourages the introduction of new marine species and microbes outside their natural habitat. For example, in southern Australia, starfish larvae native to the North Pacific, capable of remaining planktonic larvae for 180 days, were accidentally introduced via ballast water. They seriously threatened local ecosystems and even caused the extinction of an endemic species, the spotted handfish.

Treatments do exist to clean ballast water, but chlorination, electrochemistry, oxidation and ozonation, although very effective, are still chemical processes that also have an impact on the environment. How do we choose between the plague and cholera? Between invasive alien species and chemical pollution?

Finally, maritime or coastal infrastructures also significantly alter aquatic balances. Dams, lagoons, artificial barges and other marine or port activities degrade ecosystems downstream and in the surrounding marine areas. The reduction in water flow and sediment modifies water temperatures, fragments habitats and profoundly harms plankton populations.

Studies conducted around dams on the Jiulong River in China show that plankton abundance is much higher in the reservoirs than in the upstream and downstream sections.

Blocking natural flows encourages the proliferation of algae, which unbalances the ecosystem. Similarly, recent studies show that structures linked to offshore renewable energies (wind turbines, floating solar panels, tidal power plants, etc.) also disrupt ecosystems through noise, electromagnetic fields, light and hydrodynamic changes. These changes affect the stratification of the water column and the flow and circulation of nutrients.

Regulations and standards to assess the impact of human activities on plankton communities are now urgently needed.

Dead zones, the deadly trap closing in beneath the waves

Some consider them the greatest threat to the ocean, worse than global warming: oxygen-depleted areas where life dies out or disappears. A tragedy for biodiversity and our climate. The phenomenon, called 'eutrophication', begins with an influx of nutrients – mainly waste from human activities – flowing into the ocean. These nutrients stimulate the development of microalgae that absorb a large amount of oxygen. When they die, these algae sink and are decomposed by bacteria, which also consume a lot of oxygen. Simultaneously, algae on the surface prevent light from penetrating the deeper layers of the ocean. Darkness limits photosynthesis and therefore the production of oxygen. The trap closes in on the petrified area.

After that, everything accelerates. Deoxygenation kills a large number of marine animals. Their decomposition by bacteria consumes the little oxygen that remains. In these conditions, the krill and copepods also die. Their disappearance causes stagnation of the waters, which are no longer stirred up by the movements of these plankton in search of food. The result is a huge silent, inert zone, frozen in a macabre stasis.

Although some fish manage to escape to more oxygenated areas, their concentration in small areas makes them easy prey for fishing. Corals, shellfish and other crustaceans die of asphyxiation. Once petrified, they will be decomposed by bacteria, thus contributing to maintaining the phenomenon. Only jellyfish, which consume very little oxygen, can withstand the cataclysm and even thrive.

In the Baltic Sea, within a few decades, all higher life forms disappeared and were replaced by primitive bacteria. Large-scale dumping of weapons and munitions after the two world wars exacerbated the phenomenon, releasing copper, heavy metals and chemical agents that are toxic to the surrounding plankton.

In these dead zones, oxygen-producing plankton is replaced by bacteria producing nitrous oxide (N_2O), a greenhouse gas 300 times more destructive than CO_2 and the main contributor to ozone layer depletion. Today, the oceans, and in particular dead zones, represent 25 per cent of emissions of this gas, seriously contributing to the warming of our climate.

In recent years, this phenomenon has been on the rise. The UNEP (United Nations Environment Programme) counted 136 dead zones in 2004, 400 in 2008 and more than 700 currently. Since 1950, the number and extent of these zones have increased tenfold, with peaks of acceleration in the 1970s and 1990s. In 2008, a study showed that, over the last ten years, the expansion of oxygen-depleted areas represented a total of 6.6 million square kilometres, or twelve times the surface area of France.[83] The largest of these areas are off the coast of Louisiana and New York, in the Bay of Bengal and in the Arabian Sea. The dead zone in the latter alone already measures 70,000 km², almost the size of Hungary.

The tragedy of dead zones is not confined to the ocean. It is even more common in lakes and lagoons, where the phenomenon occurs even faster.

The responsibility clearly lies with humans: global warming, pollution, construction and trawling all release nutrients buried in the soil. However, we can combat this scourge if it is treated early enough and the genetic capital of the area has not been lost. In the United States, in the Chesapeake Bay drainage basin, which has been under threat since the 1950s, decreasing the use of nitrates has reduced runoff by 25 per cent and restored some of the oxygen. Life has partially returned since then.

However, if we do nothing, these dead zones will remain as huge marine cemeteries for thousands of years. Silently and invisibly, far from our cameras and social networks, a major disaster is playing out; one that threatens humanity and ecosystems. Planktonic diversity is the main character, and could help us reverse this tragedy in the coming decades. We urgently need to understand and protect this diversity, because if it dies in these waters, so does the foundation of our history and of our future.

Sardines go wild for chestnuts

Pierre Mollo, one of France's greatest plankton advocates and a wonderful storyteller, likes to humorously remind us why sardines love chestnuts. This anecdote clearly demonstrates how the management of coastal areas must take into account the effects induced by our human activities on plankton. Over the last few years French local authorities have been planting conifers on a massive scale. These fast-growing trees optimize the absorption of greenhouse gases. The initiative is commendable, but imperfect, since these trees only produce needles, unlike deciduous trees which produce humus rich in mineral salts. These nutrients are transported by rivers and reach coastal waters, where they encourage the development of photosynthetic plankton around estuaries and bays. These plankton feed zooplankton, which feed crustaceans, which in turn are eaten by small fish, including sardines. Thus, by planting conifers en masse instead of deciduous broad-leaved trees, we are depriving plankton of nutrients and jeopardizing the entire food chain, right down to sardines.

But the consequences don't end there: by limiting the development of plankton, we are reducing its capacity to sequester carbon. Also, although conifers absorb CO_2 this effect is cancelled out, or even reversed, by the decline in plankton.

In Japan, oyster farmer Shigeatsu Hatakeyama follows the same logic. In 1989, his oysters became unfit for consumption due to a tide of red algae. Shigeatsu then realized that dinoflagellates, which are toxic and inefficient at sequestering carbon, feed on agricultural pollutants carried by the nearby river. So he suggested that the villages upstream from his oyster farm should plant a forest of chestnut, oak and other deciduous broad-leaved trees. Thanks to local support, the dream is becoming a reality. Water quality is improving and the oysters are becoming edible again. Since then, Shigeatsu has become well-known for being the man who planted trees to feed his oysters, earning himself the nickname 'Grandpa Oyster'.[84]

Decades later, his initiative has set an example for others. Tens of thousands of trees have been planted along riverbanks in Japan, and pesticide reduction programmes are being implemented. Shigeatsu's lesson has also made it into school textbooks. Every year, thousands of children take part in tree planting, to raise awareness of the connection between human activities, river water, the sun, the forest and the sea.

The management of coastal forests need to better integrate their impact on coastal waters. The land can nourish the sea – or pollute it. The choice is ours.

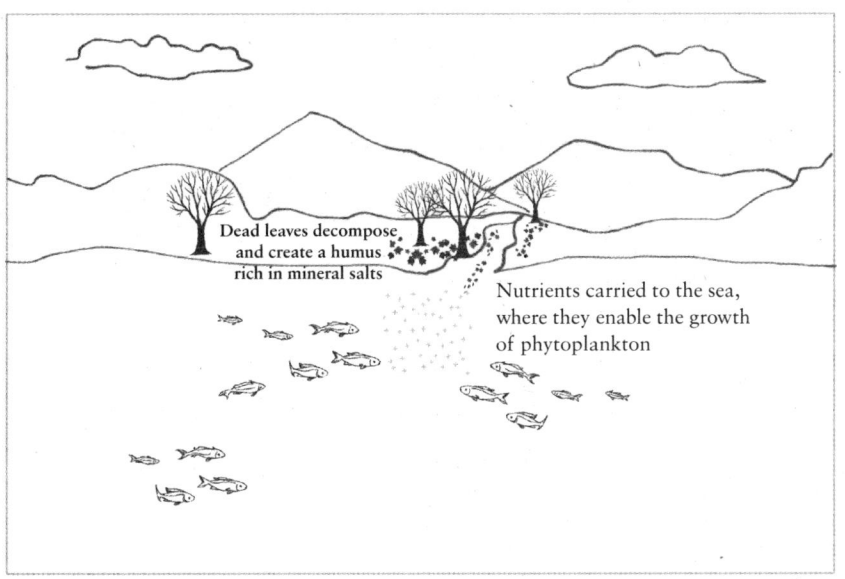

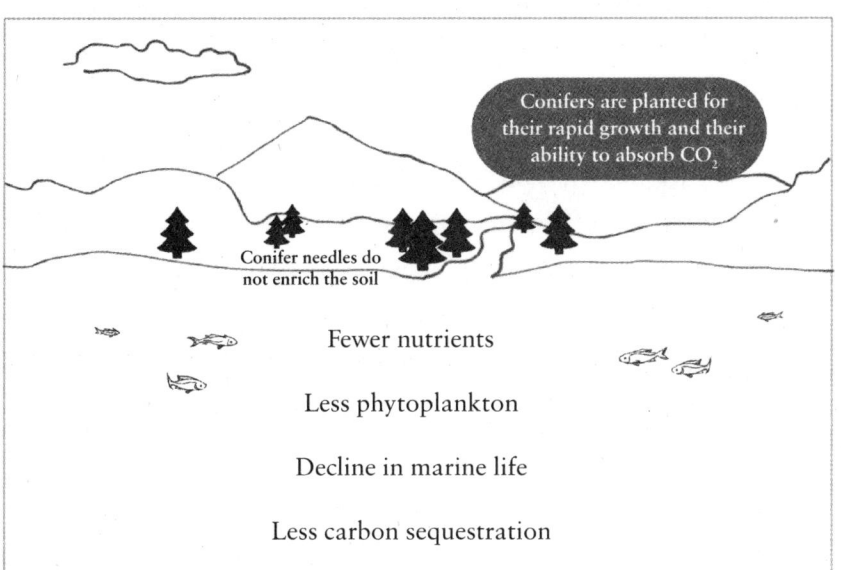

Link between terrestrial vegetation and the oceans

Toxic red algae tides
versus zombie plankton

All of the imbalances caused by human activity encourage the development of invasive and often toxic microalgae tides. These are developing in all regions of the world, invading the coastlines with scarlet dinoflagellates which, in addition to the toxins they emit, suffocate all biodiversity. These spectacular, high-density blooms can extend over more than 1,000 km², decimating everything in their path. Some years, hundreds of tonnes of corpses of dolphins, turtles and other fish wash up on beaches. In other cases, toxins are released into the air, causing damage that reaches far beyond the ocean. The southern United States and many parts of Asia are being particularly affected. Population density and coastal pollution exacerbate the phenomenon.

In the French West Indies, like in most tropical areas, ciguatera, a food poisoning caused by the dinoflagellate *Gambierdiscus toxicus,* now affects over 50,000 people a year. Mentioned in China as early as the seventh century, the Spanish conquistadors described it as early as 1606 when they landed in Vanuatu, well before this toxin decimated Captain Cook's crew in 1774 – after horrific suffering.

The toxins cause severe digestive, neurological and cardio-vascular problems. The spread of ciguatera is clearly linked to the disappearance of corals, themselves victims of climate change, which are being replaced by algae conducive to the development of this toxic plankton. Sometimes bathers are poisoned simply through contact or inhalation.

In the Bay of Bengal, blooms of *Noctiluca scintillans*, a luminescent dinoflagellate, delight tourists but threaten biodiversity. Although the microalgae itself is not toxic, it is invasive and suffocates the entire ecosystem, leading to the creation of large dead zones.

In the Mediterranean, *Ostreopsis* is one of the most dangerous in its category. Historically, it was only found in tropical areas and in moderate quantities. But warming waters have encouraged its development over the past twenty years. So much so that it is now found on the Atlantic coast of Spain and France, turning the water dark and putrid. Its toxins are released into the air, causing irritation, coughing and asthma attacks. Consuming fish or seafood that feed on these plankton poses a risk to human health. Due to thousands of cases of poisoning, the authorities were forced to close beaches for several weeks during the summer in Biarritz and in all affected areas.

Other microalgae, such as certain cyanobacteria or the marine diatom *Pseudo-nitzschia*, have a nasty habit of becoming both invasive *and* harmful. The latter, via the food chain, can lead to neurological disorders in mammals and seabirds. Alfred Hitchcock was inspired by this phenomenon to write the screenplay for his very famous film *The Birds*. In humans, this diatom causes serious memory problems and even convulsions, leading to coma. In early 2025, shellfish from northern Brittany were still banned from being sold because of this toxic diatom.

These red tides also have disastrous economic consequences: entire sectors of activity linked to the ocean, such as tourism, aquaculture and fishing, have been brought to a standstill. If these events continue, they will eventually render all seafood products unfit for consumption, and therefore unmarketable. Not to mention the impact on biodiversity. According to the Scripps Institute of oceanography in California, the financial cost to the United States of these tides of toxic microalgae is estimated at 7 billion dollars a year, a figure that is constantly rising.

Highly sophisticated monitoring systems have been put in place to control the influx of these toxic algae and anticipate the damage they could cause. Some scientists are also questioning the current acceleration of the phenomenon, attributing it to the increase in the number of checks being carried out. Either way, it is undeniable that global warming and pollution linked to human activities do play a decisive role.

There are solutions, however, some of which are in the form of zombie plankton. Recent studies show that some toxic dinoflagellates are, in certain conditions, parasitized by other smaller dinoflagellates.[85] These reproduce very quickly inside, producing up to 400 offspring that will devour the nucleus and slowly digest the chromosomes of the host organism, keeping it in a state halfway between life and death. Scientists have named them 'zombie plankton'.[86] After a while, all the parasites exit the host in a nightmarish worm-like form.

Yet the outcome is actually fairly favourable for our ecosystems. For more than twenty years, northern Brittany

– particularly the Bay of Morlaix – has thus been rid of the invasive toxic dinoflagellate blooms that tended to invade the Penzé River, following the passage of these parasites. The phenomenon is not rare, and many viruses primarily serve to regulate the growth of plankton populations.

It is therefore essential to begin to understand, recognize and protect these various drifting allies, which can help us to restore balance. Otherwise, the phenomenon of invasive algal blooms will only continue to spread.

Plastic versus glass, with plankton as referee

The use of plastic packaging rather than glass was encouraged by the chemical and agri-food industries after the Second World War for economic reasons. At the time, nobody was aware of the environmental impact of this new material. Today, despite our growing environmental awareness, major brands – foremost among them a well-known American soda company – justify their use of petroleum-based packaging by claiming it has a lower climate impact. This narrative, echoed by the major plastic producers, is explained by the fact that packaging is lighter than glass or metal, reducing transport costs and the associated carbon emissions. Plus, plastic production requires less energy than glass. But this view once again overlooks the effects of plastic on plankton.

Plastic pollution is a tragedy for the ocean. Every minute, nearly a million plastic bottles are sold around the world, and the majority will not be recycled. The lifespan of plastic in nature ranges from 450 years to infinity. Our annual production is approximately 460 million tonnes and is expected to triple by 2060 at the current rate of growth. Yet only 9 per cent of this plastic is recycled. The rest ends up in our oceans. ADEME, the French Agency for Ecological Transition, estimates that every minute, the equivalent of a lorryload of plastic is dumped into the ocean.[87]

The images of turtles and birds suffocated by plastic bags are heartbreaking, but they may well mask the reality of the most serious consequences of this tragedy, which for the most part is invisible. Because these petroleum-based products also play a crucial role in the composition and behaviour of plankton communities. Microplastics alter the photosynthesis of certain plankton and reduce the appetite of zooplankton, leading to slower growth, shorter life expectancy and a drop in reproduction. These factors all affect the entire food chain.

What's more, zooplankton often mistake these microplastics for phytoplankton. Once they've been ingested, they become part of the food chain, all the way up to the biggest fish. And therefore, they reach us. Whales, which feed mainly on krill, absorb up to 10 million microplastic particles a day. When ingested by zooplankton, these highly buoyant plastics slow down the descent of their excrement, which reduces carbon sequestration in the deep sea. The problem is also found in diatoms. They tend to cling to microplastic, which prevents them from sinking after they die.

In light of these effects on marine biodiversity, we must therefore urgently change this dogma on the supposed advantages of plastic for our climate compared to other packaging. It is now clear that petroleum-based packaging is damaging marine ecosystems, particularly planktonic ones, and cancelling out the supposed benefits. In fact, compared with plastic, glass has a far lower impact on the climate, is far more recyclable and has a much more limited impact as waste in the natural environment. The road to hell is paved with good intentions...

Life cycle	Glass	Plastic
Production	⊕	⊖
Transportation	⊕	⊖
Lifespan in nature	⊖	⊕
Recycling	⊖	⊕
Impact on marine ecosystems	⊖	⊕
Visible climate impact	⊕	⊖

⊕ high impact
⊖ medium/low impact

Life cycle analysis: the hidden impacts of plastic

The University of Plymouth is planning to launch research in this area to quantify the impacts of plastic waste on the carbon sequestration capacity of plankton. But who is willing to fund such research, when the results could undermine the biggest agri-business lobbyists?

Glass is just one of many examples. Bamboo, seaweed and cereals are also great alternatives that consume even less energy. In this fight, thanks to the consumer choices we make every day, we can all help to bring about change.

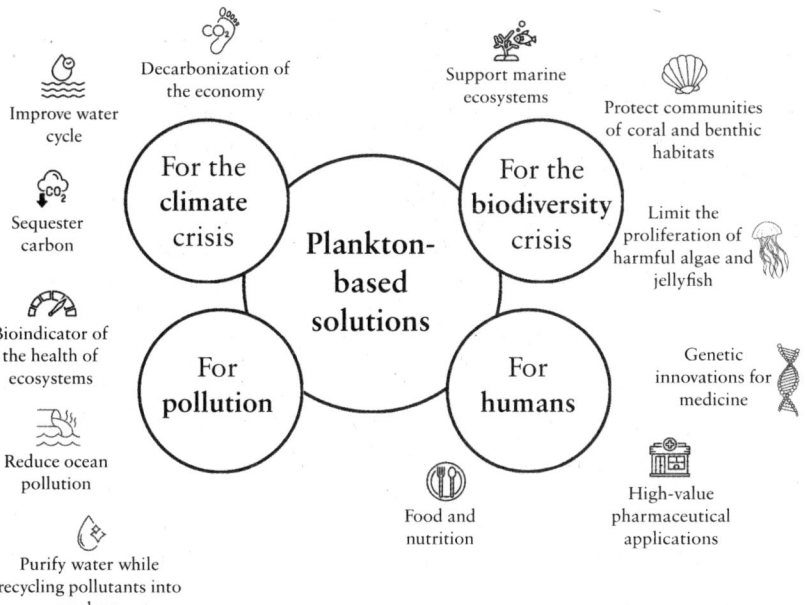

Improve water cycle

Decarbonization of the economy

Support marine ecosystems

Protect communities of coral and benthic habitats

Sequester carbon

For the climate crisis

For the biodiversity crisis

Plankton-based solutions

Limit the proliferation of harmful algae and jellyfish

Bioindicator of the health of ecosystems

For pollution

For humans

Genetic innovations for medicine

Reduce ocean pollution

Purify water while recycling pollutants into products

Food and nutrition

High-value pharmaceutical applications

Solutions, innovations, plankton!

Plankton that speak to us: from autopsies, to the health of the planet

In the early twentieth century, miners used to take a canary with them when they went down the mines. As long as the canary sang, everything was fine. But as soon as toxic gases, particularly carbon dioxide, became too strong, the bird would stop singing. The miners interpreted this silence as a warning signal and hurried back to the surface. Long before gas detectors were invented, canaries were used to warn of the invisible danger surrounding miners. Isn't plankton the canary warning us of the invisible threats to our planet?

When an invisible element becomes visible, it reveals new information. In the case of plankton, this information could allow us to anticipate the state of our ecosystems and take action before it is too late.

Due to their very short lifespan and rapid rate of reproduction, planktonic organisms react quickly to changes. This means that there is very little lag time between a change in the environment and its impact on the community. Furthermore, they drift with the currents rather than swim, which makes it easy to predict their movements. Finally, coastlines and the edges of large lakes have the advantage of being at the intersection of land, water, sediment and air, making them unique places of transition between different biotopes, and ideal areas for monitoring their evolution.

Chemical inputs from different industries create a multitude of different environments. A high diversity of plankton generally indicates a healthy ecosystem, while a decrease in abundance can signal habitat pollution or degradation, the causes of which need to be investigated.

There are countless applications for plankton, even beyond the environmental spectrum. For example, in forensic medicine, diatoms are used as markers in cases of drowning. Because they are ubiquitous, the type and quantity of them in a person's lungs will help determine when, where and how they died. If plankton can provide information about a decomposing human body, it can also tell us about the state of the living world itself, helping to prevent it from completely breaking down.

Plankton should be our 'canary in the coalmine' to alert us to the health of aquatic ecosystems and detect possible contaminations. By monitoring plankton, we can identify areas of overfishing, sudden warming, invasive algal blooms or the sudden degradation of entire ecosystems. We still face the challenge of working out how an individual population or community of plankton could be linked to disruption in our ecosystems. The primary levels of plankton are very sensitive to changes and react very quickly to modifications.

Zooplankton, such as copepods and rotifers, are highly sensitive to variations in temperature, salinity and pollution. Plankton should therefore be used as an indicator to assess possible environmental impacts before any infrastructure projects are undertaken at sea.

One of the most ambitious projects in this field is the creation of a digital twin of the ocean based on artificial intelligence. The Digital Twin Ocean[88] project faces a number of obstacles: data availability, compatibility and quality, as well as cost, and the capacity to cover the whole ocean. Although we know what type of plankton is present in an area and in what quantities, it is still hard to know how these plankton interact with each other, how they communicate, what symbioses they may form, who will eat whom, and above all how they will evolve in the future. By creating a digital twin, we can provide more advanced models, particularly for modelling climate change. These initiatives should help us to live in greater harmony with the living world and repair the damage we cause.

All these advances should inform policies and encourage the effective monitoring of major international treaties aimed at protecting the environment, especially the ocean. Plankton is the ocean's canary: it's up to us to listen out for its silence!

Plankton versus global warming: 'Give me a half tanker of iron, and I will give you an ice age'

In 2005, American entrepreneur Russ George founded Planktos Inc. with the stated ambition of restoring bio-diversity and slowing climate change via plankton. The ocean absorbs 30 per cent of our emissions, largely thanks to these microorganisms. With this in mind, Russ George proposes fertilizing the ocean with iron to stimulate the multiplication of plankton capable of sequestering carbon, and therefore cooling the planet. He actually got the idea from oceanographer John Martin, who said, 'Give me a half tanker of iron, and I will give you an ice age.'[89]

Throughout the history of life, plankton communities have successively warmed and cooled the earth, each time causing major catastrophic events and giving certain types of organism the advantage over others. In the coldest ocean regions, the growth of certain photosynthetic plankton (diatoms and coccolithophores) is limited by a lack of iron. Introducing iron into these areas would therefore promote the proliferation of this type of plankton and the absorption of CO_2. Expeditions led between 1990 and 2005 showed that adding iron increased biological activity in the surface layer. It's an inexpensive solution, but one that does nevertheless carry certain risks: can we really afford to play sorcerer's apprentice by throwing iron into the ocean to stimulate the development of plankton?

Throughout evolution, carbon has been deposited very slowly in our ocean sediments. Could a massive and too-rapid deposit of carbon risk suffocating the ocean, encouraging the development of dead zones and ultimately destabilizing this fragile ecosystem? How will marine viruses react? Will they also proliferate, in sync with other plankton? What impact will these iron additions have on water chemistry and the availability of other nutrients? In the name of the climate emergency, is anything worth a try?

One of the scientists in favour of this solution of seeding iron in the ocean once explained it to me as follows: we all know that it's dangerous to put the handbrake on when a car is travelling at 100 km per hour. But if we're driving at 100 km per hour straight into a wall and our other brakes have failed, shouldn't we try to apply the handbrake? That is precisely the situation we're in.

This seems to be the opinion shared by Russ George. In May 2007, he announced his intention to dump 100 tonnes of iron over an area of 10,000 km² off the Galapagos Islands in the Pacific, to prove the effectiveness of the concept. There was an outcry from NGOs: Sea Shepherd, already headed by Paul Watson, sent a ship to intercept the Planktos boat. Undeterred, Russ then redirected his boat towards the Canary Islands. This time, the Spanish government intervened. In response to this situation, the UN introduced a moratorium in 2008, suspending this kind of experimentation.

This did not prevent Russ George from conducting a new

illegal experiment in British Columbia. After convincing a credulous Native American community that was suffering from a declining fishery and growing unemployment, he spread 100 tonnes of iron in the Gulf of Alaska in 2012, in return for which he received the modest sum of 2 million euros from a solidarity fund supporting these populations. He promised them that this initiative would bring the fish back. To this day, it remains the largest fertilization experiment ever carried out.

While George claims that his experiment encouraged the growth of photosynthetic plankton over an area of around 16,000 km², quadrupling the production of salmon in the vicinity and encouraging carbon sequestration, others claim that no correlation can be made between that year's record fishing quota and the iron dumping, adding that the experiment above all encouraged the proliferation of the toxic diatom *Pseudo-nitzschia*. The Canadian government eventually filed a complaint against the entrepreneur, who had taken advantage of a legal loophole in international ocean governance to escape without serious consequences.

Today, many researchers, including in Europe, are calling for the moratorium on testing to be relaxed, citing the climate emergency and the technological progress made over the last two decades. While the first priority is to reduce greenhouse gas emissions, we all know that moderation alone is not the way to reverse the trend towards global warming. The most obvious solution is to create carbon sinks for storing greenhouse gases. Three major ecosystems already fulfil this role by each absorbing a third of the emissions: soils, forests and oceans. Optimizing them is crucial in the fight against climate change. In the case of the ocean, atmospheric carbon enters the water column where it follows two very distinct trajectories: the solubility pump based on currents, and the biological pump driven by life.

The first is a physio-chemical process based on the circulation of large bodies of water, regulated by temperature and salinity. These currents carry carbon from the surface to the seabed where it can remain for centuries before being released into the atmosphere. Although it captures about 80 per cent of the carbon in the oceans and provides sustenance for deep-sea organisms, on a geological scale, this system does not sequester carbon on a long-term basis.

The biological pump, on the other hand, transforms carbon into living matter through photosynthesis. In this equation, plankton represents 99 per cent of the process. Other photosynthetic marine plants, such as macroalgae and mangroves, play only a minor role. Our microscopic drifters use carbon from surface waters to create their organs. When they die, their heavy shells sink as marine snow to the bottom of the ocean, burying carbon for thousands, if not millions, of years. This mechanism represents only 10 to 20 per cent of oceanic CO_2, but is essential for climate cooling. By consuming carbon from surface waters, plankton maintains a differential that promotes the continuous absorption of atmospheric CO_2. This second option is of interest to Russ George because it can be valued in dollars.

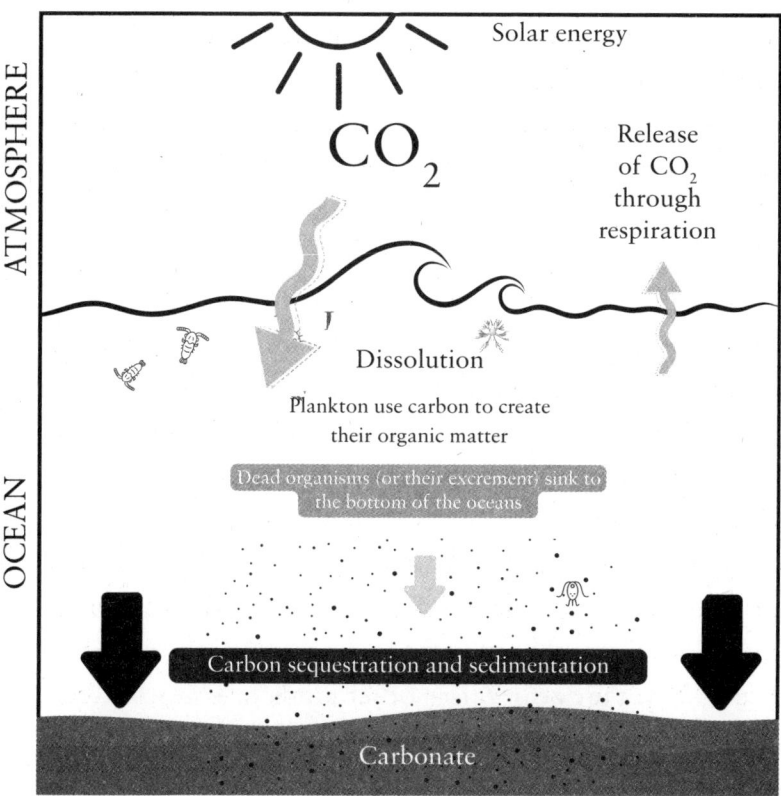

Simplified diagram of the biological carbon pump

Indeed, the businessman has often pointed out that, without plankton, the carbon cycle in the ocean would be much less efficient, or even almost non-existent, and that this mechanism has weakened over the last few decades with the decline of certain plankton populations. To the extent that, one day, we might end up with an ocean that actually rejects carbon instead of absorbing it. This extreme would signify the almost immediate end of a large proportion of the species on our planet, including our own, regardless of any austerity measures taken elsewhere. Climate stabilization therefore depends on preserving this biological pump and the plankton communities most efficient at sequestering carbon.

But behind these pseudo-ecological objectives and the veneer of his former collaboration with Greenpeace, the businessman has above all developed a potentially very lucrative business model: accumulating carbon credits via plankton. And many entrepreneurs in other countries are gaining enthusiasm for this solution. While on the face of it the project may seem appealing, in reality it is a very risky solution for ecosystems.

Other avenues are emerging, however, and some researchers are suggesting selecting viruses to regulate plankton populations. But the idea of releasing genetically modified viruses into the ocean raises other kinds of concerns. The recent Covid-19 crisis showed just how dangerous it can be to meddle with infectious agents. So how do we decide between prudent passivity and risky interventionism? We need to guard against the whims of certain entrepreneurs who are taking advantage of the climate crisis to sell so-called 'green' solutions, but we must not shy away from scientific progress either. For the ocean and for plankton, it's time to decide.

Can whale dung help to cool the planet?

The fertilization of the ocean by nutrients is directly tied to the interactions between the land and marine environments. Rainwater carries away a large number of nutrients which flow into rivers and then into the seas. Preserving terrestrial ecosystems is therefore crucial to maintaining these flows and supporting the planktonic communities that create this climate. But this fertilization also takes place in the heart of the ocean. Firstly, thanks to hydrothermal springs: these large chimneys in the abyss release nutrients from the bowels of the earth into the water. Life cycles also play an important role: the bigger the organism, the greater its impact.

Because of their size, whales and sperm whales are key to this circulation of nutrients. By hunting krill in the depths, cetaceans accumulate enormous quantities of zooplankton, which they return to the surface through their excrement, rich in iron and nitrogen. These faeces feed photosynthetic plankton, which are essential for capturing CO_2. A 1 per cent increase in its productivity would allow it to capture hundreds of millions of tonnes of CO_2 per year, which is the equivalent of two billion trees.[90] These microorganisms then feed the entire food chain of molluscs, fish, shellfish and other seafood, therefore promoting carbon absorption. Whale dung therefore plays a vital role in the proper circulation of nutrients from the depths of the ocean to the surface.

Moreover, at the end of their lives, the enormous carcasses of whales eventually sink to the bottom of the ocean, leading to the rapid sequestering of tonnes of organic carbon in marine sediments.[91]

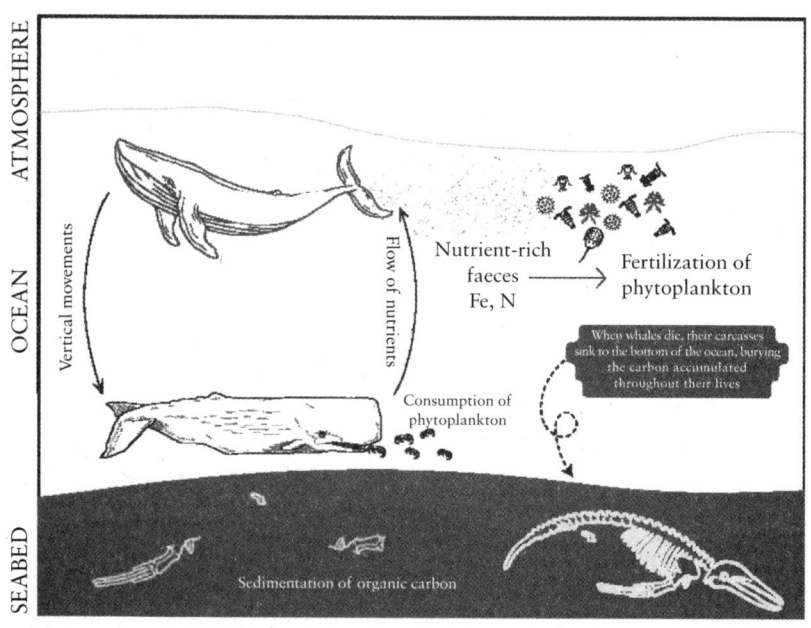

ATMOSPHERE

OCEAN

SEABED

Vertical movements

Flow of nutrients

Nutrient-rich
faeces ⟶ Fertilization of
Fe, N   phytoplankton

When whales die, their carcasses
sink to the bottom of the ocean, burying
the carbon accumulated
throughout their lives

Consumption of
phytoplankton

Sedimentation of organic carbon

Phytoplankton fertilization and carbon sequestration:
the ecosystem role of whales, the Whale Pump

Unfortunately, in the early twentieth century, millions of whales were hunted for their oil, meat and other products, decimating populations by 70 to 90 per cent compared with historic levels. Blue whales, for example, which numbered around 250,000 before this slaughter, now number only 10,000.[92] Human activities are also reducing the plankton populations that feed whales, causing large numbers of them to die.

As well as the threat of extinction of certain species, the depletion of certain plankton has consequences for the entire marine ecosystem and for our climate. This is why we need an integrated approach, to take action at all levels, from small plankton to large cetaceans, because every link in the chain is interconnected.

Regenerating ocean biodiversity

Plankton is essential for restoring biodiversity in the ocean. Having destroyed 50 per cent of ocean ecosystems in the last fifty years, we cannot simply be satisfied with 'sustainable development'. We need to move on to the regeneration stage. On land and at sea, if we want to restore the chain of life, we must start with its foundations. In the ocean, this is plankton.

Our invisible little diatoms struggle to attract the attention of humans, who are likely to find saving whales and dolphins more appealing. But the survival of these charismatic species depends on the plankton that feeds them. Famine is a bigger threat to them than overfishing.

Although sometimes criticized, fishing is still essential for billions of human beings, particularly in the poorest countries. According to the Food and Agriculture Organization of the United Nations (FAO), aquatic foods provide at least 20 per cent of the daily animal protein intake of more than 3 billion people worldwide.[93] Seafood products therefore contribute to food security – particularly in terms of protein intake – and to the generation of income. However, while consumption of these marine products has risen over the last thirty years, fish stocks have plummeted.

Increasingly sophisticated industrial fishing technologies and higher demand are putting unbearable pressure on these wild resources. The FAO estimates that one third of the world's fish stocks are overexploited, seriously threatening many species whose renewal depends on plankton. Yet current fisheries management is still focused on catches and fails to consider the underlying ecosystems.

Since the 1990s, the United Nations has imposed limits on the estimation of a species in relation to fishing figures, because this conventional approach does not include a comprehensive view of what enables the development of these fish stocks upstream.

Countless studies demonstrate the direct link between planktonic primary production and the productivity of fisheries in given areas.[94]

It is now essential to adopt an Ecosystem Approach to Fisheries (EAF)[95] that integrates planktonic activity, rather than limiting ourselves to the evolution of a few species that have significant commercial value or visibility. This requires a better understanding of planktonic systems and data which are currently insufficient, especially in the open sea.

The current crisis is forcing fisheries management organizations (FMOS) to reform without delay. The challenge is to understand the ocean system as a whole and our 'canary plankton' is crucial here.

Global warming – and the disruptions that result from it – means that plankton populations are being forced to move towards the poles to find more temperate waters, thereby altering the entire food chain. Some fish cannot follow such a migration and are under threat of extinction. Cod from the English Channel, for example, have migrated to Norway due to lack of food, financially ruining local fishermen who had followed the recommendations regarding quotas.

Sardines in Brittany have recently seen a decrease in size and weight, to the extent that they are no longer fit for sale. In 2024, Concarneau's purse seiners (fishing vessels with large nets) returned with empty holds, due to a lack of catches the right weight or age to be sold. In the Mediterranean or the Bay of Biscay, the situation is even more stark: the weight of sardines has halved in twenty years. Between 2006 and 2016, in the Mediterranean, the available biomass of sardines fell by about two thirds, from 200,000 to less than 67,000 tonnes.[96] The size of individuals has followed exactly the same trend, while their life expectancy is a seventh of what it used to be. This is an ecological, economic and social tragedy. As a result, French sardines have now virtually disappeared from fishmongers' stalls and a large proportion of the sardines we buy in Europe are imported from Morocco.

Studies have shown that these changes in sardine metabolism are directly linked to plankton. As global warming has reduced the size of photosynthetic plankton, this phenomenon has had repercussions for zooplankton. These smaller zooplankton require more physical effort from the sardines. They are reduced to filtering water instead of hunting. This expenditure of energy prevents them from accumulating enough fats and nutrients (especially omega-3).[97]

These imbalances are also causing collateral damage to other fish, dolphins and birds. As a result, some sardine fishermen have been forced to switch to fishing hake. This puts additional pressure on an already overexploited fish population. Sometimes the consequences are just as dramatic, but harder to identify. For example, the huge increase in accidental dolphin fishing is directly linked to the reduction in plankton size. The lack of plankton means that sardines have had to expand their predation zone and move closer to the coast, where there is more net fishing. They are followed by the dolphins that feed on them. In coastal areas, they end up trapped in fishing nets intended for other, more abundant species.

As for whales in the North Pacific, while it was hoped that their numbers would soar after the whaling ban came in, they have actually fallen by 20 per cent between 2012 and 2021. This was the result of a rise in water temperature of 4 to 10°C above normal levels. This variation has led to a decline in the plankton communities that feed the krill, and so on. More than 7,000 whales have starved to death for this reason.[98] In North America, the hunting of grey whales, which was very widespread in the 1970s, was gradually banned. However, in 2019, seventy starving whales were beached between Alaska and California.[99] Terrible images, but they mask an even more tragic reality, because the majority of these mammals are dying at the bottom of the ocean, out of sight. Once again, the best way to save whales is not (only) to tackle the boats that hunt them, but to protect the invisible, microscopic pasture they feed on.

Even seabirds are suffering as a result of these changes: by trying to follow the fish they usually feed on, which are moving to colder seas, they leave their nests for too long. This move is fatal for their offspring.[100] The heatwave that hit the North Pacific between 2014 and 2016 caused the disappearance of several million seabirds.[101] Due to a lack of plankton, the fish that these birds feed on have been unable to feed and reproduce in the usual proportions.

Over the last sixty years, 70 per cent of seabirds have disappeared. The entire balance of life on earth is threatened by changes within the ocean's plankton communities. The influence of plankton is therefore not purely confined to the oceans. These drifting, mobile microorganisms connect a whole range of biotopes and enable life to develop throughout the planet.

We must save the corals!

Besides fish, there are many other species that depend directly on plankton. Corals, natural works of art with their delicate shapes, cover less than 0.2 per cent of the ocean floor but are home to more than 30 per cent of currently known marine species.[102] In these lush gardens anchored in the seas, an area of around one square kilometre contains as much biodiversity as the whole of mainland France.[103]

Saving the coral reefs is therefore essential to preserving marine biodiversity. And once again, to do this, we need to understand the role of plankton. Corals live in symbiosis with colourful microalgae called zooxanthellae, which they host in exchange for food. If they are subjected to thermal stress, corals expel the zooxanthellae, become bleached and then die of starvation. Understanding these planktonic interactions in detail is critical. Other planktonic microorganisms play a vital role, such as the planktonic bacteria that supply corals with vitamin B, while others help to reduce turbidity and water pollution. Some carbon-absorbing plankton reduce ocean acidity, which weakens coral skeletons and makes them more vulnerable to predators, waves and pathogens floating in the ocean.

It is therefore clear that some solutions can be found in plankton. There are many factors that encourage plankton biodiversity: temperature, currents, salinity, light, nutrients and human activity. A better understanding of these mechanisms would make it possible to restore ocean biodiversity. In Australia, it has recently been shown that dying coral reefs can be saved by collecting planktonic microbiota from around healthy corals and transplanting them onto the diseased corals. This coral microbiome transplantation (CMT) technique is similar to human faecal microbiota transplants. The health of corals and the ocean also depends on their microbial balance.

While global planktonic biomass is not decreasing overall, its composition is changing at an exponential rate and the species that depend on it no longer have time to adapt, causing a dramatic domino effect: extinction is hitting cetaceans, birds, fish and terrestrial mammals, one after the other. The human race might be the next victim.

Marine protected areas
for plankton?

Today, just as the Climate COP brings together all stake-holders fighting against global warming, the Biodiversity COP brings together those fighting for both marine and terrestrial biodiversity. It is astonishing – and frankly worrying – to see the extent to which, even in these dis-cussions, the biodiversity of the ocean remains largely ignored. And even more so, the invisible life forms within the ocean.

As a result of these international gatherings, ecologically and biologically significant areas (EBSAs) were defined for land, in order to facilitate the establishment of nature reserves essential to the living world. Unfortunately, it took a few more years before they considered defining such zones for the world's largest connected ecosystem and the source of all life on earth: the ocean.

Correcting this oversight made it possible to define the famous marine protected areas (MPAs)[104], the establish-ment of which has been so hotly debated throughout the world. The assessment criteria for these zones were based on those used for terrestrial systems. Undoubtedly not the best decision from a scientific point of view.

It will therefore come as no real surprise to learn that these international negotiations on global biodiversity, and in particular marine biodiversity, simply do not include the issue of plankton, the foundation of all biodiversity.

In 2024, at the sixteenth COP on Biodiversity (COP16) held in Cali, Colombia,[105] they confirmed the commitment to protect at least 30 per cent of coastal and marine areas by 2030 in the form of marine protected areas, and to restore 30 per cent of degraded ecosystems. Of the hundreds of presentations given during the two-week conference, only one was partially devoted to plankton. Plankton is not mentioned anywhere in the agreements. Further discussions in Rome in 2025 to finalize the agreement on financing the actions did not do any better. When it comes to defining the areas to be protected on the planet, and particularly in the oceans, no one is trying to understand which areas are priorities for the historical and functional basis of our biodiversity: plankton.

This is all the more shocking given that today's technology makes it possible to map plankton activity and the ecosystem services plankton provide around the world. Oceanic landscapes could be mapped out in the same way as on land, with deserts, prairies, steppes, mountains or valleys. After all, the ocean is an incredibly complex three-dimensional space containing 'regional' specificities such as depth, nature of sediments, luminosity, nutrients, temperature, the presence of seamounts or currents that can reveal the dynamics of life that differ greatly from one area to another.

Used properly, this map would indicate the priority areas for action to encourage planktonic activity. These actions could therefore have a knock-on effect on all marine and even terrestrial ecosystems, as we explored previously. Actions that support these ecosystems, such as limiting external disturbances and reintroducing endemic species that have disappeared from these planktonic ecosystems, are lacking. They would, however, be crucial in restoring the health and biodiversity of the ecosystems that depend on them.

There are only a handful of visionaries, with a clear understanding of how life works in the ocean, who are advocating for the establishment of 'key oceanic planktonic areas' (KOPAs) that take this part of the living world seriously.

Discussions are just beginning. It's a long road ahead.

Plankton for cleaning our waste

Have you ever wondered where the clean water that flows from your taps comes from? Once again, plankton plays an essential role.

With freshwater becoming increasingly scarce, using plankton to purify wastewater is a critical issue for tomorrow's world. Our drinking water is in danger of becoming desperately scarce, and it is essential to develop more sustainable treatments to purify it. Especially as the abundance of nutrients contained in all our different kinds of waste is both a resource for the future of a world in short supply, and a major challenge for our societies.

Various types of plankton have long been used in wastewater treatment plants in addition to other more mechanical or chemical technologies. Natural and energy-efficient, certain microalgae are capable of producing valuable resources from our pollutants. They offer a promising alternative to our current processes that are very costly and emit greenhouse gases.

In a wastewater treatment plant, the primary treatment is essentially mechanical: the largest residues are isolated by decantation, filtering and other sedimentations; and then the secondary treatment eliminates residues dissolved in dirty water using suspended microorganisms, including plankton. A better grasp of these organisms could make these processes more efficient, more eco-friendly and less expensive.

Filtration and decantation
Removal of the largest residues

Microorganisms (plankton)
Degradation of organic contaminants

Absorption of dissolved nutrients
Microalgae and cyanobacteria absorb nitrogen and phosphorus

Management of excess microalgae
Zooplankton consumes the surplus microalgae

Recycling of by-products
Potential production of biogas, fertilizers and animal feed.

Purified water ready for consumption or sustainable disposal

Wastewater purification cycle

Microalgae and cyanobacteria absorb nitrogen and phosphorus, which are elements from domestic, industrial and agricultural pollution. Limiting the concentration of these substances before they are released back into nature is imperative, because they encourage the proliferation of algae, increase the risk of eutrophication and contribute to widespread toxic tides which have devastating consequences for ecosystems.

We know, however, that phosphate reserves, mainly extracted in Western Sahara under conditions close to slavery, are likely to run out quickly.[106] This will considerably reduce global agricultural production and put many countries at risk of food shortages. Some scientists are already talking about a 'phosphogeddon' with consequences even more terrible than climate change in the coming decades.[107] A lack of phosphate resources would cause significant drops in yields and widespread famine. Needless to say, by then the prices of the resource will have increased and will have had repercussions on basic food products. This is why microalgae could prove to be a valuable tool for recovering nitrates and phosphates, and recycling them before they are lost in wastewater. All over the world, 'new generation' wastewater treatment plants are aiming for an energy-positive, carbon-negative model with a recoverable plankton biomass. This feat could be achieved through the use of microalgae working in synergy with bacteria.

This idea is not new: tests were carried out as early as 1960, but our lack of expertise when it comes to these microorganisms prevented us from going any further. Today, advances in genomics and AI are making the invisible not only visible, but also valuable. In France, the start-up company NXO Engineering, based in Montpellier, is developing an 'algae bioremediation' concept by optimizing the symbiosis between bacteria and microalgae.[108] The resulting agal biomass is recycled through methanation. The company Zeni (short for 'zero nitrate') in Saint-Nazaire targets food processing and intensive aquaculture industries to treat their water loaded with nitrates and phosphates.[109] At the end of the cycle, phytoplankton biomass is recycled in the form of bio-based fertilizers or biostimulants for more sustainable agriculture. In the Bordeaux region, the company Carbon Works recycles the CO_2 from cement plants to produce microalgae for use in energy, food processing and cosmetics. A fine example of industrial ecology, where waste from one industry becomes a resource for another.

In the long term, some even envisage using zooplankton to consume excess microalgae, which could then be recycled in the form of animal feed, pharmaceutical products or to restore degraded ecosystems.

In the words of the French chemist Antoine-Laurent de Lavoisier, 'Nothing is lost, nothing is created, everything is transformed.'

From Chernobyl to cleaning up our system

This plankton-based sanitation system could even extend to rivers, seas or the ocean. Plankton-based 'bioremediation' is generating growing interest around the world. In 2024, a Franco-Argentine team shed light on the mechanisms that enable a species of microalgae to absorb excess zinc from rivers.[110] The concept could be applied to other heavy metals to clean up lakes, rivers, ponds and even the ocean. Some ports and marinas are already exploring ways of absorbing heavy metals using cyanobacteria and microalgae. Others are looking into recycling the water used to clean boats. Finally, photosynthetic plankton, capable of converting carbon into biomass, could prove invaluable in oil-producing coastal regions.

It could also help tackle plastic pollution. In 2016, Japanese scientists discovered a planktonic bacterium (*Donellea skaiensis*) secreting enzymes similar to those of certain marine fungi that have demonstrated their ability to degrade polymers in some of the most commonly used types of plastic (especially PET).

Plankton even deals with radioactive waste. In 1986, after the Chernobyl disaster, salps (gelatinous plankton) helped limit the exposure of marine species by filtering large quantities of water and ingesting the radioactive particles present in plankton and seawater before sinking to the depths after their death. These plankton have therefore managed to achieve a kind of natural decontamination by transferring radioactivity from surface waters to deep sediments. The same phenomenon was observed after the Fukushima accident in 2011.

The potential of plankton is not limited to the aquatic environment. In northern France, start-up Bioteos is developing an air pollution control system based on microalgae grown in glass tubes filled with water.[111] These algae absorb fine particles of suspended pollutants (carbon monoxide, ozone, nitrogen dioxide, sulphur dioxide). An initial large-scale trial has been launched to purify 700 m3 of air per hour in the Lille metro system. The company now offers a range of products for cleaning the air in rooms in offices, hotel lobbies and even private homes. Once again, the harvested microalgae are repurposed as agricultural biostimulants. In a similar vein, in Singapore, the company Urban Greening is integrating microalgae walls to purify the air and produce sustainable fertilizers. Similar initiatives are emerging in Europe, in France, the UK, Germany and Belgium.

At the industrial level, there are still many challenges to be met for scaling up, in particular the very high cost of these processes. Photobioreactors require constant lighting, which reduces their profitability. In addition, optimizing plankton strains through genetic selection will be a crucial factor in improving the efficiency of treatments and the reuse of by-products. Further research and innovation are definitely needed, but plankton-based waste management solutions offer considerable potential for an eco-friendly and truly regenerative future.

Ángel León, the 'chef of the sea': plankton broth and bioluminescent dishes

In the 1980s, Ángel León was a child who preferred to spend his days fishing in the Bay of Cádiz rather than going to school. When he wasn't gazing out to sea at this port – which saw the departure of the great expeditions of the conquistadors to America in the sixteenth century – he was busy cooking. One day, he dreamed of combining his two passions: the sea and gastronomy. In 2007, the young chef opened his first restaurant, Aponiente, in a small street in El Puerto de Santa María on the Atlantic coast of Andalusia.[112] His ambition was to develop luxury gastronomy based solely on seafood products, with the lowest possible environmental impact. He believes that the sea is 'the garden of the future'.[113] When he started out, his efforts were met with mixed success. As his debts mounted, he gradually resigned himself to giving up his restaurant and putting it up for sale. As a final challenge, in 2008, he came up with the idea of incorporating microalgae into his cooking. He wanted to be the first chef in the world to put plankton on the menu.[114]

While looking for an aquaculture farm, he stumbled upon one of the few companies producing microalgae-based fish feed. Expecting it to be located on the other side of

the world, he sent an email in English asking for information. The director, Carlos Unamunzaga, invited him to
visit his modest industrial site. Ángel then discovered that
Fitoplancton Marino is also based in El Puerto de Santa
María, just a few hundred metres from his restaurant. Fate
brought these two destinies together, marking the start of
a collaboration that has become legendary in the world of
gastronomy.

The chef is a visionary who wanted to use plankton in his
cooking, to make luxury dishes that are delicious, incredibly nutritious and have a very low environmental footprint.
But how would he find the best plankton for eating? There
are so many of them. Rising to the challenge, Carlos began
cultivating dozens of different plankton. The two friends
tested the taste and texture of different microalgae under
various growing conditions. After two years of experimentation, they finally selected a purified microalga with
an incomparable taste. Angel decided to use it as the basis
for his Plancton Marino[115], a delicious broth that gives his
dishes the inimitable flavour of the sea without requiring a
large quantity of fish or seafood.

Four years later, Carlos and Ángel had not only finalized
the production and preparation methods for their plankton,
they had also obtained authorization to sell this algae on
the European food market. In 2014, the product was finally
launched, and was an instant success. Ángel was awarded
a second Michelin star at the end of that year, and a third
in 2017. In the meantime, he moved his small restaurant to
a huge nineteenth-century building by the sea, at the tip of
Santa María. A regular guest on the TV programme *Top*

Chef in several countries, Ángel is now known as the 'Chef of the Sea'.

As for Carlos's company, it now mass-exports what has become their only product to twenty-five countries that innovate using this unusual microalgae. Some people even make nougat with it.

But Carlos and Ángel didn't stop there. Fascinated by the properties of plankton, they came up with the idea of using luminescent dinoflagellates to illuminate dishes served in restaurants. With the flick of a spoon, the plates light up in the dark! The experience is unique, and Netflix has seized on the story to produce a feature-length documentary tracing Ángel's journey. In 2019, his restaurant was ranked among the world's best 100 restaurants and he is ranked thirteenth in the global list that recognises the 100 most outstanding chefs in the world, The Best Chef Award 2023.

Today, Carlos and Ángel are continuing their research and are discreetly working away on a diatom that tastes surprisingly similar to certain meats. From moray eel mochi, rice made from eelgrass (an aquatic plant) and mussel sausage, they are continually inventing new recipes. Thanks to these pioneers, microalgae can now be used by a wider audience as a condiment and ingredient.

In France, Algama is now offering a microalgae paste that can replace eggs in custards, cakes and pancakes.[116] Its nutritional value is excellent (it is very rich in omega-3 and 6, vitamins C and E and minerals), plus cultivated seaweed has less to complain about than battery hens, and both its taste and its binding properties are close to those obtained with traditional eggs.

Ángel is continually researching ways to make seafood products eco-friendlier and more flavourful. And when some customers point out the incredible fishy flavour of the dishes enhanced with his Plancton Marino product, Ángel corrects them: plankton doesn't taste like fish, fish tastes like plankton!

Spirulina, the superfood of the Aztecs and astronauts

Despite the hard work of a few pioneers such as Ángel and Carlos, there is only one type of plankton that is truly famous throughout the world: spirulina. This cyanobacterium, which lives suspended in water, gets its name from its filamentous, spiral shape. In recent years, when talking to people about macroalgae, I've often heard the following remark: 'I already eat seaweed, I take spirulina!' This plankton that we took to be green algae is actually a blue bacteria. For marketing reasons, the EU and other bodies accepted the inaccurate term 'microalgae' to refer to this type of plankton (deeming "bacteria" to be a hard sell from a marketing standpoint...). Considering our vast ignorance of the underwater world, this linguistic approximation is not too consequential in the grand scheme of things. Cyanobacteria have now officially been recognized as microalgae, so we'll use that term.

Lexicological debates aside, it has to be said that spirulina, or *Arthrospira*, has a lot going for it. We've been using it since ancient times: the Mayans are said to have cultivated it in their canals several thousand years ago, and some believe that its decline brought about the fall of their civilization in the ninth century. The Aztecs called it *tecuitlatl*, which roughly translates to 'stone's excrement'.

They harvested this slimy material from freshwater lakes, dried it into cakes and consumed it as a source of protein and nutrients. In fact, Francisco López de Gómara wrote in his book *General History of the Indies* that the Aztecs ate it the way we eat cheese.

Emperor Montezuma, a great lover of fish, served it to his servants to increase their endurance so that they would bring back their precious catch as quickly as possible. Hence their nickname of 'fish runners'. Today, without realizing it, champions such as Messi, Ronaldo and Benzema still follow this example and use spirulina to improve their performance. Unfortunately, the conquistadors came in search of gold, not bacteria that smells like cheese. The draining of the salt lakes that followed the colonization of Mexico precipitated the disappearance of spirulina. Production was not resumed until 1970, in Lake Texcoco.

In the Republic of Chad, too, spirulina has been consumed since the ninth century. Considered beneficial and protective, it is primarily reserved for pregnant women and newborn babies. These days, the descendants of the prestigious Kanem-Bornu Empire, which dominated the region for 700 years, continue to take a few grams of the precious cyanobacteria every day. It wasn't until 1967 that Belgian botanist Jean Léonard demonstrated the nutritional benefits of this 'superfood'. In fact, Brussels now has an 'Avenue de la Spiruline' in its honour!

Spirulina has an outstanding nutritional profile, with a protein content of around 60 per cent, which varies depending on environmental conditions and production methods.[117] By way of comparison, soya, the basis of animal feed, contains only 25 per cent. As well as quantity, it also has quality going for it. The amino acid profile of these proteins is extremely interesting and corresponds almost exactly to the FAO recommendations in terms of daily intake.[118] But its benefits aren't limited to protein. Also rich in iron, vitamins, minerals and trace elements, it has unrivalled anti-oxidant and anti-inflammatory properties and strengthens the immune system.[119] This unique composition makes it an essential food for improving endurance, reducing muscle fatigue and promoting recovery after intense exercise.[120]

Recent studies show that spirulina improves heart health by reducing cholesterol and blood pressure.[121] [122] [123] It is also thought to have extremely promising antibacterial and antiviral properties, and even anti-cancer properties. Finally, it could mitigate the side effects of chemotherapy and radiotherapy by reducing cell oxidation. Other recent studies suggest that it reduces the symptoms of depression and anxiety by supporting the balance of neurotransmitters in the brain, detoxifying the body and eliminating heavy metals (mercury and arsenic) and other toxins.[124] [125]

An increasing number of humanitarian organizations, including the Red Cross, are now turning to this superfood to help combat malnutrition in disadvantaged populations. The WHO declared in 1996 that spirulina was the 'food of the future', and UNESCO defined it as 'the ideal and most complete food of tomorrow'.[126] These cyanobacteria, now being considered to feed astronauts, as we saw earlier, were among the first living organisms and a food consumed by humans since antiquity. Truly a 'blue gold'!

Chlorella almost fed the world in the post-war period

As impressive as the benefits of spirulina are, its nutritional profile is not an exception in the world of plankton. Still within the field of microalgae, there are many other organisms that have great potential for our health. According to the European Commission, only three species of plankton are currently recognized as food for humans: the cyanobacterium spirulina, the diatom *Odontella aurita* and the green alga *Chlorella*, which is now widely marketed as a dietary supplement.

Chlorella had its heyday in the 1940s. After the war, the Americans turned to this plankton – which had been used by the Germans during the First World War – to meet the food needs of their population. Renowned for its extraordinary levels of protein, essential fatty acids, as well as vitamin B12 – a rare and critical compound for our brain that is otherwise only found in meat – *Chlorella* was a godsend in that period of shortage. In 1950, a large farm was set up in Stanford, California, with the objective of producing 10,000 tonnes of protein per year over an area of 4 km^2 and employing only twenty workers. A leading scientific magazine announced that tomorrow's world would be free of famine thanks to the cultivation of microalgae harvested from pond scum. Unfortunately, being right too soon is the same as being wrong. Production costs proved to be astronomical. At the same time, traditional agriculture was benefiting from the incredible innovations – both mechanical and chemical – that had emerged during the war. As a result, *Chlorella* cultivation was quickly shelved. Today, it is exclusively cultivated for the food supplement industry in the form of tablets, powders and liquid extracts. It is also increasingly being studied as a biostimulant for sustainable agriculture or for decontaminating nutrient-saturated environments. In early 2025, Coca-Cola's main distributor in Israel announced the opening of a 2,500 m^2 plant producing a high-protein drink using *Chlorella*.[127] The goal was to offer a vegan and low-tech alternative to dairy products. The result is a white powder naturally extracted from *Chlorella* after fermentation, and the final drink contains 70 per cent protein and has a profile rich in amino acids. It is colour of milk and doesn't taste like seaweed but has a fairly neutral flavour. Many companies are undertaking similar projects. More than a century after its initial uses in the trenches of Verdun, *Chlorella*'s time may have come.

Health-boosting microalgae

Odontella aurita is a small marine diatom with an astonishing structure and a very distinct elongated shape. Like its fellow creatures, it produces a silica shell, but it is above all its contribution of marine omega-3 (EPA and DHA) that makes it so valuable. Essential to the functioning of our brain, eyes and nervous system, these fatty acids also play a key role in combating obesity, diabetes, cancer, arthritis, asthma and cardiovascular diseases. Without these omega-3s, our brain cannot function and our neuronal activity is greatly reduced. Given these benefits, it will come as no surprise that the price of these molecules reaches almost 5,000 euros per kilo and that the commercialization of these little diatoms is now authorized by the European Union.

Another substance derived from microalgae, astaxanthin, is developing rapidly in the pharmaceutical industry. This compound, which has an antioxidant power 6,000 times greater than that of vitamin C, is generating a lot of interest. In the human body, astaxanthin has the unique ability to cross the blood-brain barrier to act directly in the brain, improving its performance by limiting neuroinflammation. The powers of this red pigment are also recognized for their highly effective protection of the skin and eyes. In the United States, high-tech farms have been selling them since 1999.

There are numerous innovations based on microalgae around the world. In the Czech Republic, one start-up is using them to grow synthetic pork in the laboratory. The concept is to produce real pork, without slaughtering or causing suffering to any animals, using stem cells that grow in tubes to produce meat. Growth serums used to make cells reproduce are usually based on bovine foetal cells. But in this case, these serums will be based on microalgae, rich in proteins and vitamins.

Other companies are even using microalgae to produce food supplements by recycling industrial waste. MiAlgae Ltd in Scotland, for example, offers omega-3 pills made from seaweed fed on waste from whiskey distilleries.[128] In 2024, the start-up was a finalist for the Earthshot Prize, a prestigious global environmental award supported by Prince William. MiAlgae claims to produce thirty times more omega-3 than fish. In Singapore, a project using microalgae to absorb waste from soya residues and sugar or beer production is also underway.

Interest in the properties of microalgae is nothing new. Even in ancient times, Pliny the Elder made reference to a 'flower of salt, that yields a sort of oily fat'.[129] This description suggests the presence of a 'salty oil' on the surface of salt marshes. However, it could be a natural secretion from lipid-rich microalgae. Other texts reveal that saline mud was, at that time, prized for its medicinal or cosmetic uses. Today, these microalgae appeal not only to Michelin-starred chefs but also to gamers. Often too consumed with their gaming to care about their diet, gamers are now consuming plankton to improve their cognitive performance and stay focused for long hours in front of their screens. In addition, the antioxidants in certain microalgae protect their eyes from the damage caused by blue light. Finally, omega-3, magnesium and iron help to reduce tiredness and stay alert for longer.

Low-tech agriculture:
cultivating blue gold

Microalgae cultivation in general is completely low-tech: it requires little water, consumes no energy, recycles nutrients and absorbs large quantities of CO_2 while producing oxygen. It generates three times less greenhouse gas than soya, while consuming three times less water and seventy-five times less agricultural land. Finally, the nutritional health benefits of these microalgae are extraordinary.

Currently, Europe mainly imports its soya from South America, a production that is mostly based on GMOs and devastating for the environment, as 70 per cent of the soya that the EU buys today to feed its cattle comes from Brazil and has contributed to the deforestation of the Amazon. Not only would microalgae have a much better yield, but they would also make it possible to relocate protein production to Europe.

In addition, certain microalgae are also very rich in lipids. According to IFREMER, microalgae cultivation could generate up to 60,000 litres of oil per hectare per year, compared with 6,000 litres for palm oil. So they offer yet another solution to combat the intensive oil palm plantations that are so destructive to flora and fauna.

Yet microalgae production remains a nascent market worth around €3 billion, with 15,000 tonnes produced annually, half of which is spirulina. The next question is how to cultivate these nutritional nuggets. Microalgae production is essentially based on two models: either open ponds or raceways; or photobioreactors, large transparent glass tubes subjected to very specific light and temperature conditions to encourage optimum development of the microorganisms. These are more modern and efficient but are also more expensive.

Growing spirulina is relatively simple: sunlight, alkaline water and heat are the ideal conditions for it to develop rapidly. It multiplies rapidly in mineral-rich lakes as soon as the temperature rises above 30°C. Its cultivation is developing worldwide, with China, India, the United States, Thailand and Mexico leading the way in production. In China, the giant BGG has developed its largest site on the border with Laos, where 2,000 kilometres of glass tubes are filled with cyanobacteria.

However, overall production remains modest at 7,000 tonnes per year, more than half of which comes from China. In fact, the country has declared cyanobacteria a 'food of national interest'. India is following the trend, with the emergence of major private stakeholders. In France, a militant model has developed with 'low-tech' micro-operations, short circuits and reduced volumes. This is referred to as 'farm-grown spirulina'. Producing 150 tonnes of spirulina annually, France is Europe's leading producer, well ahead of Spain and Italy. But its low volumes force it to import 90 per cent of its consumption.

Innovations in spirulina production continue to develop in tropical countries. Thanks to the high temperatures in these parts of the world, the mass of this photosynthetic bacteria can double every twenty-four hours.

Naturally, challenges arise, such as environmental hazards, installation costs and scaling up. The risks of farm contamination are also significant. If a single microalgae-eating protist manages to enter a tube, it will feel like we're watching *Pac-Man* on an old arcade video game. The protist gobbles up everything in sight, multiplies rapidly, and can devour everything produced in a matter of days. Therefore, despite the high retail prices, the profitability of these new production sites is often still a precarious balance.

Some people dream of one day growing plankton at sea. Just imagine drones capable of recognizing and harvesting microalgae from lakes or the sea. WWF recently announced that 10 million tonnes of invasive microalgae were available on the surface of the Baltic Sea. That's a huge amount of valuable micronutrients that we may one day be able to harvest. One way or another, these microorganisms could certainly revolutionize our food and our industry. Our recent mastery of genomics combined with the exceptional reproductive speed of microalgae will enable us to select increasingly productive and robust strains. Plus, out of the millions of microalgae in existence, only a few tens of thousands are currently categorized and fewer than twenty are being cultivated. So researchers have plenty to keep them busy for a few more years.

Once the advantages of microalgae have been recognized, will it be possible to build a sustainable and ethical model that is resistant to the excesses of capitalism seen in other sectors? It's not about what we grow, but how we grow it.

We need to draw inspiration from the history of agriculture to avoid repeating the same mistakes (such as GM crops and intensive livestock farming). Plankton farming is still currently in its infancy and so there are no competing lobbies to slow down or oppose its eco-conscious development. We are therefore looking at a unique opportunity to create a new ethical model combining biotechnology and ecology.

A move towards
zooplankton farming

In 1956, the famous marine biologist Alister Hardy noted in his book *The Open Sea: The World of Plankton*, 'It is perhaps sacrilege to mention that I have hauled up sufficient Euchaeta not only to serve the needs of science but also to provide a delicious addition to the supper table. Boiled in seawater for a moment, strained and then fried in butter and served on toast, Euchaeta is a delicacy which one day might support a small fishery to supply a luxury market.'[130]

Yet the farming of zooplankton is still non-existent. As with microalgae, there are scientific and technical barriers to achieving this ambition. Mainly because we have poor control over the production of their food, namely microalgae. With the exception of the microcrustacean *Artemia salina*, which forms the basic diet of farmed fish larvae, and a few early attempts with copepods, we have no idea how to feed and therefore cultivate zooplankton.

So that leaves fishing! Krill fishing is well established. This small marine crustacean has the advantage of reproducing very quickly and containing omega-3 fatty acids and astaxanthin, making it a highly sought-after species, and therefore a very lucrative market. Krill oil is already a luxury product with nutritional values that have been recognized for several generations in Norway and a number of other developed countries.

Sustainable fishing to ensure the proper reproduction of stocks would be perfectly feasible. Feeding people directly through krill fishing would be much more environmentally friendly than using it to feed farmed salmon. Plus, krill are among the first links in the marine food chain, so has the advantage of accumulating far fewer pollutants and heavy metals than the fish at higher trophic levels.

Worldwide, there are very few examples of plankton-based foods other than microalgae and krill. In Japan and China, dried jellyfish, marinated and cut into strips, are eaten as a festive dish. They are even farmed. These gelatinous foods are 98 per cent water, so you need to eat quite a few before you're full. And only a few species from Asia are actually edible.

In reality, the majority of microalgae and krill-based products are not produced for human consumption, but for the aquaculture production of crustaceans, fish and other molluscs. The omega-3s are extracted for the highly profitable nutraceuticals market, while the rest of the biomass is used to feed fish farms. Given the drastic decline in fish stocks and the rapidly booming global population, we urgently need to turn to new forms of sustainable aquaculture. But to do this, it is absolutely essential to master the keystone of the food chain, namely plankton.

Biomimicry: what if plankton could light our cities?

The biomimetic potential of the aquatic world, and plankton in particular, is far beyond our imagination. While the glass industry is extremely polluting and energy-intensive due to the melting of silica sand at temperatures close to 1,700°C, diatoms naturally produce ultra-resistant glass structures in water at a temperature below 10°C and without any carbon emissions. Definitely food for thought.

These marine glass creatures make optical researchers green with envy because the incredibly sophisticated geometry of their crystals allows them to filter and direct light in very small quantities. These light-capture techniques are used to improve the efficiency of photovoltaic panels.

Diatoms are also capable of secreting a very powerful adhesive that allows them to withstand the most violent pressure and currents. An asset that is already inspiring the automotive and aerospace industries to strengthen structural assemblies. These marine adhesives can also be used to develop self-cleaning coatings. They are based on a complex polymer structure produced by microorganisms that can attach themselves to a surface and form an antiparasitic layer. By organizing itself into a biofilm, this gel acts as a protective barrier, releasing chemical substances that prevent undesirable organisms from sticking to it. This natural film could be particularly useful for protecting surfaces that are exposed to bad weather or humidity, such as boat hulls, maritime infrastructures or even buildings constructed in very rainy areas. It offers an eco-friendly solution that could replace conventional chemical treatments which are often harmful to the environment.

Another promising potential of microalgae: polar diatoms and certain cyanobacteria produce antifreeze molecules capable of lowering the freezing point of water. This could provide a sustainable alternative to chemical and toxic antifreeze.

Others naturally produce energy resources by synthesizing hydrogen from water using specific enzymes. Unlike conventional methods of hydrogen production such as steam methane reforming, or water electrolysis, this plankton does not emit CO_2 and is not dependent on any fossil resources. This is a very promising method but one that still needs to be improved because current yields and costs mean it is not very competitive compared with other fuels.

Plankton also have the ability to recover microparticles of precious metals (gold, silver, palladium) from the water, by integrating and accumulating them within their cells.

In terms of colouring, spirulina produces a water-soluble pigment that is very effective in giving drinks, dishes and cosmetics an emerald-blue colour. It can be found on packaging under the barbaric code 'E18', which does little to honour the only natural colourant of this colour.

Plankton also offer solutions for replacing plastic. Some planktonic bacteria and microalgae naturally produce PHA (polyhydroxyalkanoate), a polymer which they store as granules inside their cells when they are exposed to an excess of nutrients and a shortage of nitrogen or phosphorus. This is a mechanism similar to fat storage in animals. This polymer can be extracted and transformed into bioplastics that are biodegradable in water and soil. This bio-packaging can then be used in medicine (sutures, implants) because it is highly compatible with human cells. Produced from organic waste, it is also water-resistant, making it possible to manufacture films, packaging and other everyday objects (toys, combs, glasses, etc.).[131]

Even zooplankton are inspiring new technologies: in underwater robotics, for example, the speed and efficiency of copepod movements are used as a model for improving naval propulsion systems and the movement of aquatic drones.

Last but not least, we should mention Glowee,[132] a company exploring the bioluminescent potential of certain plankton to illuminate our cities and offer an alternative to increasingly expensive and energy-intensive electric lighting. A fascinating idea, especially since these planktonic light bulbs emit a soothing, gentle light that is much less oppressive than traditional bulbs. Imagine a future where all our homes are lit by diatom glass bulbs filled with glowing dinoflagellates.

In 4.2 billion years of evolution, plankton has already revealed extraordinary abilities. We *Homo sapiens* have only been around for 300,000 years, so we can learn a lot from these more experienced, ingenious marine organisms.

Cosmetics, health, fertilizers and energy: so much more to discover!

Plankton has already inspired numerous innovations, starting with cosmetics. Most anti-ageing creams, rejuvenating serums, moisturising masks, shampoos, soaps and after-sun care products contain them, because these microalgae are bursting with vitamins, minerals, antioxidants and essential fatty acids. Since the 1940s, the Biotherm brand has established itself as a pioneer with its Life Plankton range, now a global benchmark.

Applications in the healthcare sector are equally promising. While some compounds found in microalgae inhibit the proliferation of cancer cells, others, such as the phycocyanin in spirulina, or the chlorophyll and phytosterols in *Chlorella*, have anti-inflammatory properties.

In Spain, researchers are injecting microalgae under the skin to produce oxygen and regenerate the epidermal surface. Seaweed has long been used for grafts and dressings. Even jellyfish collagen – similar to our own – is frequently used in dermatological treatments and grafts, because it is more effective and durable than bovine or porcine collagens. Some molecules are also known for treatments related to the digestive system, arthritis, hypertension, joint pain, ulcers...

In 2008, research into fluorescent proteins from jellyfish – which enable cancer cells to be located ten times more effectively than a scanner – was awarded the Nobel Prize in Physics. Even krill, which have an exoskeleton made of chitin, a substance found in our tissues, is proving to be a valuable medical resource for reducing allergic reactions in skin and bone grafts.

If plankton is good for our health, it's not surprising that it's also good for other living things. In agriculture, it offers plants an alternative to chemical fertilizers and pesticides. In ancient times, the Romans placed jellyfish at the base of grapevines to hydrate and fertilize the soil. At the mouth of the Mississippi river in the United States, disruptions caused by widespread use of pesticides have led to such a proliferation of jellyfish that some fish farmers have given up shrimp farming in favour of selling the invasive jellyfish in the form of biostimulants.

Certain microalgae are rich in phosphorus, potassium and nitrogen, and therefore protect and improve the soil and groundwater, boost plant resistance and promote water retention. Thanks to their numerous biological and nutritional properties, planktonic fertilizers could facilitate the transition from agriculture based on chemical inputs to a more sustainable form of agriculture, while improving yields by 30 per cent and reducing the amount of water required to grow crops.

Thanks to millions of years of experience fighting viruses, bacteria and fungi in the ocean environment, plankton acts as a natural pesticide in the terrestrial environment, limiting invasions of fungi and insects. Given that the use of (mostly petroleum-based) pesticides has almost doubled globally between 1990 and 2018,[133] with dramatic consequences for the health of soils, the ocean and humans (cancer, infertility, poisoning...), it's imperative that we find eco-friendly alternatives, and soon. In 2016, the French company ImmunRise identified a microalgae that can combat the mildew that has plagued European crops since the nineteenth century. This fungus was responsible for the destruction of potato crops in Ireland, leading to the Great Famine that occurred between 1845 and 1849. Today, some forms of downy mildew have evolved to become resistant to the fungicides traditionally used to control these organisms. But none of the various types of mildew and other fungi affecting crops have ever come into contact with marine molecules. They have therefore not developed any resistance to them. Winegrowers – mildew's main victims – will undoubtedly be queueing up to find out more about this new solution.

Since producing this type of microalgae requires a large amount of heat, in 2023 ImmunRise developed a bold partnership with an Icelandic company, Algalíf. Together they created the 'Seeds of Lava' project which involves using geothermal energy of volcanoes as a natural, inexpensive and carbon-free resource.[134]

Photosynthetic plankton also has the potential to revolutionize the energy sector. After all, fossil fuels are just sedimented plankton. With its incredibly high renewal rates, plankton has created large fossil reserves on the ocean floor throughout the past 4 billion years. Since these plankton carcasses or excrement are composed of hydrogen and carbon, the result of their sedimentation were logically named 'hydrocarbons'.

When they reach certain temperature and pressure

conditions underground, these organic deposits transform into gas or oil. Simply put, if our plankton reach a depth of 60°C on earth, they form gas. Higher up in the strata if they find the necessary conditions in the rock, they sometimes form oil. In both cases, these formations are the result of an incredible series of geological circumstances which leads to a strange paradox: after cooling the planet by sinking to the bottom of the ocean, plankton transform into hydrocarbons, which then power our car engines, simultaneously warming the planet's atmosphere.

As Paul Falkowski, an American specialist in phytoplankton puts it, 'We have been using oil from fossil phytoplankton to fuel our cars and heat our homes for more than a century. Each year, we burn oil that took a million years to produce...' [135] The next time you fill up with petrol, perhaps you'll think of those millennia of marine life poured into the tank of your car and consumed in just a few kilometres.

If dead plankton can be transformed into fossil fuel, could we not consider taking advantage of similar metamorphoses in living plankton?

Not only do these organisms reproduce incredibly fast, but their exceptional lipid content makes them a credible alternative to traditional biofuels. In their natural state, plankton can store up to 50 per cent lipids by dry weight (compared to 40 per cent for rapeseed, or sunflower[136]). When grown under stress conditions (nitrate deficiency, light intensity), some species increase their lipid production until it reaches 80 per cent of their dry mass. Rates such as these are unique in the living world, and make it possible to produce high yields of biodiesel. We could also consider producing bioethanol by fermenting the sugars present in certain species of algae; or producing biogas, or even hydrogen, by letting these plants break down in oxygen-free (anaerobic) environments.

In 2010, EADS, the parent company of Airbus, demonstrated that an aircraft could fly on fuel derived from microalgae. According to the company, only 100 kg of algal raw materials were needed to produce 22 litres of algal oil, or 21 litres of refined fuel. Maybe one day we'll be able to fly without creating a carbon footprint, because 100 kg of algae will absorb 182 kg of CO_2 while they are growing. Microalgae production does not require fertile soils and does not compete with other terrestrial food resources. They can also absorb carbon from cities and large polluting industries. Unfortunately, once again, our inadequate scientific knowledge of these organisms, combined with the production costs and technological challenges, have forced even the most optimistic to put their algae-fuelled planes back in the hangar, probably for a few more years.

Towards a 'neocolonialism' of plankton resources?

Plankton could become a major geopolitical issue intersecting food security, climate change and energy independence. As a largely unexplored genetic reservoir, it is now attracting the interest of a very small number of countries.

Historically, the UK has long been a pioneer in this field, notably with Plymouth's Continuous Plankton Recorder (CPR Survey), the only global tool for measuring plankton, and an unrivalled source of data for over eighty years.[137] Thanks to the CNRS in Villefranche-sur-Mer, near Nice, has taken advantage of its exceptional location on the Mediterranean to become a world-renowned centre for plankton. This laboratory has been at the heart of the scientific development of the *Tara* expeditions, a unique, extraordinary initiative that has boosted our knowledge of plankton and of marine ecosystems.

Outside of Europe, the United States and Japan dominate research into marine microorganisms, while Norway and Chile are acquiring expertise, especially with regard to their aquaculture activities. Australia, under the aegis of UNESCO, is developing its own measurement tools, with a new version of the continuous plankton recorder. But above all, it is China that is investing massively in this area.

Studying plankton requires advanced technologies and costly infrastructures (special boats, stations, marine scientists, etc.). The Nagoya Protocol may well aim for fair and equitable sharing of benefits arising out of the utilization of genetic resources, but for the time being, only a few wealthy countries actually have the capacity to collect and exploit this data.

Furthermore, the 'high seas', which represent approximately 65 per cent of the oceans, are hard to govern as they are outside of national jurisdictions. Strategic or high-intensity areas for plankton such as the Arctic, Antarctic and the South China Sea are also often areas with difficult access and tense geopolitical situations, making cooperation even more complicated.

Since the first patent on a marine genetic resource in 1988, the number of patents has boomed. Almost 13,000 genes were patented in 2028.[138] Microorganisms alone – mostly planktonic – account for 73 per cent of these patents. In the coming years, there is a major risk of seeing the development of a kind of neocolonialism of marine genetic resources, with a dozen countries and a few companies monopolising the majority of the knowledge and licenses needed to exploit the phenomenal potential of this little-known resource.

The situation is already pretty concerning. By 2018, the German company BASF, the world's largest chemicals company, held 47 per cent of all patented marine resources, surpassing the share of the 220 other companies combined (37 per cent). Universities and their public academic partners had only registered 12 per cent of these patents by that date. Basically, only ten countries in the world hold 98 per cent of plankton patents. Germany, with BASF, accounts for almost half, and if we add the United States and Japan, that is almost 75 per cent of the patents held by just three countries. Which leaves 165 other countries with no patents on these marine genes, even though some of them have vast coastlines.

To regulate this hoarding of knowledge and resources, after eighteen years of hard-fought negotiations, in 2023 the UN adopted a treaty on the high seas (BBNJ *Biodiversity Beyond National Jurisdiction*), which entered into force in January 2026. The stakes are high. The treaty aims to better understand the impact of our activities on the marine environment and to better protect sensitive areas by creating marine protected areas. Finally, the convention guarantees the equitable sharing of marine genetic resources and the expansion of training initiatives in the least developed countries. In order for this ambitious treaty to be implemented, it needed to be ratified by at least sixty member countries. France was the first country in the European Union to do so. In September 2025, over two years after the treaty was signed, sixty-one countries finally ratified it, which will hopefully set us on the path to implementation.

Despite some progress, control of these marine genetic resources therefore remains a cause for concern. Securing a truly binding agreement towards the collective and responsible management of the world's greatest shared biodiversity, needs to become a priority for everyone. These new areas of research will dominate in the years to come, and the political world urgently needs to address these issues. Ensuring multilateral, ethical and equitable development is crucial to preserving and sharing these resources that are essential to the future of life on earth.

Conclusion

Plankton is everywhere on our planet. It is in freshwater, in saltwater, in ice, and even in the air. It creates a living link between continents and their ecosystems, connecting seas and coasts, streams and valleys, atmosphere and marine sediments, subsoil and clouds. Its immense genetic diversity and multiple forms connect it to all areas of the living world, of which it is the cornerstone. Over the course of earth's history, it has shaped living conditions in the ocean and then on the land, making our planet hospitable and playing a key role in regulating our climate and balancing our ecosystems.

The great biologists James Lovelock and Lynn Margulis understood this when they formulated the Gaia hypothesis in the 1970s. Their theory is that the biosphere functions like a gigantic organism fed by the sun's energy. Within this community, each component self-regulates and contributes to maintaining conditions that are favourable to life. From the outset, plankton was central to the researchers' theory, which was was initially based on the observation that when the temperature rises, plankton can regulate the climate by releasing a sulphurous gas (DMS) that encourages the formation of low, white clouds all around our planet, thus mitigating its warming. An immense preservation mechanism comparable to that of a planetary immune system.

In a way, as human beings, we are already, at our own level, microscopic 'Gaia' made up of several billion eukaryotic cells, bacteria, archaea, fungi and other viruses that cooperate to ensure our survival. Is the same true for planet earth? Could Gaia be a gigantic symbiosis drifting through space? A planetary cell? In this immense holobiont, every species, every organism, every molecule is just a tiny part that evolves within a larger entity that is still unknown to us. Perhaps one day we will realize that this set of ecosystems has even ended up developing a form of consciousness.

The late philosopher Bruno Latour sums up the hypothesis as follows, 'The earth is a totality of living beings and materials that were made together, that cannot live apart, and from which humans can't extract themselves.'[139] This Gaia theory clashes with the Darwinian evolutionary view that organisms adapt to their environment. In Gaia, it's the organisms that define the environment: a fundamental proposition which, in terms of significance, rivals Galileo's discoveries. The environment revolves around life, like the earth revolves around the sun. Not the other way round! The Gaia hypothesis therefore contains the beginnings of a way of thinking that may become the common sense of a civilization yet to come.

If Gaia exists, then plankton, drifting freely and reproducing at an unparalleled speed, connects all its elements.

By capturing CO_2, producing oxygen, supporting the base of the food chain and ensuring the functioning of the biogeochemical systems that regulate our planet, plankton is the great architect that maintains the conditions for life. This basic premise is confirmed by sedimentary archives, which show the extent to which significant declines or shifts in plankton communities are markers of major environmental and climatic changes throughout the history of our planet. So the earth does not follow a fixed trajectory but oscillates around variable equilibria, allowing life to continue in one form or another.

Plankton invisibly connects every part of the living world. And yet, like an elusive shadow, plankton sheds light on the depths of our ignorance. Our lack of knowledge can viewed as an immense field of possibilities, a door opening to new discoveries, and a source of optimism for saving our planet.

3 billion years ago, invisible marine bacteria transformed carbon into oxygen and edible matter. Why not use these same tools to repair our ailing planet? Transforming atmospheric carbon to produce food and oxygen on a large scale could be an extraordinary answer to the two biggest problems of our time: climate change and world hunger. Isn't this what spirulina growers are already doing as they install vats filled with emerald cyanobacteria on the roofs of sun- and carbon-saturated cities to produce a highly nutritious resource?

Even if plankton doesn't have the solution to all our problems, it does offer us a positive outlook. New lifestyles, biological and technological innovations are possible. Plankton invites us to imagine a world where collaboration takes precedence over competition, where nature and science come together to design a more sensitive, beautiful and harmonious society. By revealing its planktonic mysteries to us, the living world is opening

up a new path, one that is full of promise for generations to come.

Acknowledgements

Many thanks to Chris Bowler (CNRS Researcher at the Institute of Biology of the École Normale Supérieure and Scientific Director of Tara Oceans), to Fabien Lombard (Lecturer specialising in zooplankton at the CNRS Laboratory in Villefranche-sur-Mer) for all their help in reading and editing this book.

Thanks to Mascha Canaux, Bettina Meyer, Ogata Hiroyuki, Lionel Guidi, Kevin Smith, André Abreu, Ron Milo, Maris Stulgis, Abigail McQuatters-Gollop, Gustaaf Hallegraeff, Paul Falkowski, Flora Vincent, Carlos Duarte, Kevin Flynn, Andy Knoll, Dai Minhan, Camila Fernández and the eminent plankton specialists who shared their experiences and stories on the subject with me. This book is above all a collective work, the product of their research and expertise.

Thanks to Alix Le Corre for her invaluable help with the research, illustrations and bibliography of this book.

Thanks to Christian, Fiona, Lyndy, Katrina and the Legend Times team, as well as Erik, Jean-Stéphane and the Lloyd's Register Foundation family for supporting this project from the beginning.

Notes

1   Le phytoplancton, un monde méconnu qui fait vivre la planète. https://www.mnhn.fr/fr/le-phytoplancton-un-monde-meconnu-qui-fait-vivre-la-planete.

2   *Terra Cultura* https://terra-cultura.com/le-flamant-rose-le-plus-connu-des-echassiers-mediterraneens/ (2024).

3   Zhang, X., Wan, H., Jin, M., Huang, L. & Zhang, X. Environmental viromes reveal global virosphere of deep-sea sediment RNA viruses. *J. Adv. Res.* 56, 87–102 (2024).

4   https://planktonplanet.org/wp-content/uploads/2025/02/Plankton-Planet-PlanktonManifesto_MG_DIGITAL-2.pdf.

5   Six, C., Ratin, M., Marie, D. & Corre, E. Évolution de la photosynthèse dans les océans: le cas de la cyanobactérie *Synechococcus. Proc. Natl. Acad. Sci.* 118, e2111300118 (2021).

6   Duporge, F.-X., Étienne, J., Gabrié, C. & Fakir, V. *Les écosystèmes marins dans la régulation du climat* https://horizon.documentation.ird.fr/exl-doc/pleins_textes/2025-04/010093714.pdf (2015).

7   Bopp, Laurent, Bowler, Chris, Guidi, Lionel, Karsenti, Éric, & De Vargas, Colomban. *L'océan, pompe à carbone (Fiche scientifique).* https://www.ocean-climate.org/wp-content/uploads/2017/02/ocean- pompe-carbone_FichesScientifiques_04-2.pdf.

8   Rôle chimique du phytoplancton. Océanopolis https://www.
plancton-du-monde.org/module-formation/diato_04.html
(2011).

9   FAO. *The State of World Fisheries and Aquaculture 2022.*
http://www.fao.org/documents/card/en/c/cc0461en (2022)
doi: 10.4060/cc0461en.

10   Verne, J. *Vingt mille lieues sous les mers* (Librairie Générale
Française, Paris, 1870).

11   UN Global Compact, O. S. C. *The Plankton Manifesto - A
Call for Plankton-Based Solutions to Address The Triple
Planetary Crisis (Biodiversity, Climate & Pollution).* https://
ungc-communi cations-assets.s3.amazonaws.com/docs/
publications/Plankton Manifesto_MG_DIGITAL-2.pdf
(2024).

12   L'Hadéen: dans le berceau de la Terre | MNHN. https://
www.mnhn.fr/fr/l-hadeen-dans-le-berceau-de-la-terre.

13   Moody, E. R. R. et al. The nature of the last universal
common ancestor and its impact on the early earth system.
Nat. Ecol. Evol. 8, 1654–1666 (2024).

14   information@eso.org. Astronomers find missing link for
water in the Solar System. *www.eso.org* https://www.eso.
org/public/news/eso2302/.

15   Jayaraman, K. S. Still clueless about origin of life. *Nat.
India* (2008) doi: 10.1038/nindia.2008.340.

16   Jayaraman, K. S. Still clueless about origin of life. *Nat.
India* (2008) doi: 10.1038/nindia.2008.340.

17   Science News Staff. Setting the Embryological Record
Straight. https:// www.science.org/content/article/setting-
embryological-record-straight (1997).

18   Debourdeau, A. Aux origines de la pensée écologique: Ernst

Haeckel, du naturalisme à la philosophie de l'Oikos. *Rev. Fr. Hist. Idées Polit.* 44, 33–62 (2016).

19 CNRS, Plankton Chronicles: https://planktonchronicles. org/en/portfolio/animalvegetal-symbiosis-in-plankton/.

20 Field, C. B., Behrenfeld, M. J., Randerson, J. T. & Falkowski, P. Primary production of the biosphere: integrating terrestrial and oceanic components. *Science* 281, 237–240 (1998).

21 Bachy, C. & Baudoux, A.-C. Diversité et importance écologique des virus dans le milieu marin. *médecine/sciences* 38, 1008–1015 (2022).

22 Jacquet, S. Les virus marins sont-ils les régulateurs de l'écosystème et du climat océanique ? *Subaqua* (2017).

23 David Lewis, We, the Navigators: The Ancient Art of Landfinding in the Pacific (Australian National University Press, 1972).

24 *Traditional Ecological Knowledge Concepts and Cases* (F. Berkes et al.) https://www.researchgate.net/publication/269576226_Traditional_Ecological_Knowledge_Concepts_and_Cases.

25 *Fisheries in the Pacific* (CRIOBE / OpenEdition) https:// books.openedition.org/pacific/450?lang=en.

26 *Traditional knowledge for climate resilience in the Pacific* (PD Nunn, 2024).

27 Margulis, L., Sagan, D., Amézieux, C., Blanc, G. & Beer, A. D. *Microcosmos: 4 milliards d'années de symbiose terrestre* (Wildproject, Marseille, 2022).

28 Symbiose. *Centre scientifique de Monaco* https://www.centrescientifique.mc/fr/article/biologie-marine-fr/symbiose.

29 Symbiox | Symbiotic Biomedicine. *Symbiox | Symbiotic Biomedicine* https://www.symbiox.org/.

30 Le bloom, ce phénomène qui colore nos eaux – Observatoire du Planc- ton. https://www.observatoire-plancton.fr/le-bloom-ce-phenome- qui-colore-nos-eaux/ (2022).

31 Haematococcus pluviales Flotow 1844 | MNHN. https://www.mnhn.fr/fr/haematococcus-pluviales-flotow-1844.

32 @NatGeoFrance. De mystérieux microbes teintent le Groenland de rose, accélérant la fonte des glaces. *National Geographic* https://www.nationalgeographic.fr/environnement/de-mysterieux-microbes- teintent-le-groenland-de-rose-accelerant-la-fonte-des-glaces (2018).

33 Langley, L. How bioluminescence works in nature. *National Geographic* https://www.nationalgeographic.com/animals/article/bioluminescence-animals-ocean-glowing (2019).

34 Bioluminescence: light in the dark – Natural History Museum https://www.nhm.ac.uk/discover/what-is-bioluminescence.html?utm_source=google&utm_campaign=news&utm_medium=grants&gad_source=1.

35 Bioluminescence | NatureWeb.com. https://natureweb.com/poste/40-bioluminescence.

36 Gribaldo, S. Lokiarchaeon, un chaînon manquant entre archées et eucaryotes ? *Pourlascience.fr* https://www.pourlascience.fr/sd/evolution/lokiarchaeon-un-chainon-manquant-entre-archees-et-eucaryotes-9347.php (2016).

37 Génermont, J. Chapitre 4. Comment, dans un lointain passé, la sexualité est-elle née ? *Sci. Philos.* 87–110 (2014).

38 Rosier, F. Les diatomées, joyaux des océans et sources d'oxygène (2025).

39 Fritz, J.-P. Les diatomées, ces microalgues qui pourraient être un allié miracle contre la crise climatique. *Le Nouvel Obs* https://www.nouvelobs.com/sciences/20210514.OBS44021/

les-diatomees-ces-microalgues-qui-pourraient-etre-un-allie-
miracle-contre-la-crise-climatique.html (2021).

40 Falkowski, P. G. & Raven, J. A. *Aquatic Photosynthesis |
Princeton University Press*. (2007).

41 Hopes, A. & Mock, T. Diatoms: glass-dwelling dynamos.
https://microbiologysociety.org/publication/past-issues/
real-superheroes/article/diatoms-glass-dwelling-dynamos.
html (2014).

42 Étienne, I. Grâce à la neige marine, les océans du globe
stockeraient beaucoup plus de dioxyde de carbone atmos-
phérique que prévu! *Science et vie* (2024).

43 ENS Lyon. Les diatomées, bio-indicatrices de la qualité des
cours d'eau. — Site des ressources d'ACCES pour enseigner
les Sciences de la Vie et de la Terre. https://acces.ens-lyon.
fr/acces/thematiques/biodiversite/dossiers-thematiques/
biosurveillance-et-bioindicateurs/les-diatomees-bio-indi-
catrices-de-la-qualite-des-cours-d2019eau (2025).

44 AFP, S. et A. avec. L'odeur serait la boussole des oiseaux
au milieu des océans. *Sciences et Avenir* https://www.sci-
encesetavenir.fr/animaux/l-odeur-serait-la-boussole-des-
oiseaux-au-milieu-des-oceans_100945 (2015).

45 Bossé, C. Le rôle de l'océan: producteur d'oxygène et
régulateur du climat. *Blutopia* https://blutopia.org/ocean-
oxygene-climat/ (2020).

46 Catling, D. C. & Claire, M. W. How Earth's atmosphere
evolved to an oxic state: A status report. *Earth Planet.
Sci. Lett.* 237, 1–20 (2005) https://www.sciencedirect.com/
science/article/abs/pii/S0012821X05003924.

47 Sacleux, A. Découverte: la vie multicellulaire existait déjà
il y a 2,1 milliards d'années. *National Geographic* https://

www.nationalgeographic.fr/sciences/decouverte-la-vie-multicellulaire-existait-deja-il-y-a-21-milliards-dannees (2019).

48 Turner, J. T. The Importance of Small Planktonic Copepods and Their Roles in Pelagic Marine Food Webs. *Zool. Stud.* (2004).

49 Kiørboe, T., Andersen, A., Langlois, V. J. & Jakobsen, H. H. Unsteady motion: escape jumps in planktonic copepods, their kinematics and energetics. *J. R. Soc. Interface* 7, 1591–1602 (2010).

50 Gemmell, B. J., Jiang, H., Strickler, J. R. & Buskey, E. J. Plankton reach new heights in effort to avoid predators. *Proc. R. Soc. B Biol. Sci.* 279, 2786–2792 (2012).

51 Svetlichny, L., Larsen, P. S. & Kiørboe, T. Kinematic and Dynamic Scaling of Copepod Swimming. *Fluids* 5, (2020).

52 Svetlichny, L., Larsen, P. S. & Kiørboe, T. Swim and fly: escape strategy in neustonic and planktonic copepods. *J. Exp. Biol.* 221, jeb167262 (2018).

53 Chapuis, H. La plus grande migration animale révélée par un laser depuis l'espace. *Sciences et Avenir* https://www.sciencesetavenir.fr/ animaux/animaux-marins/la-plus-grande-migration-animale-reve- lee-depuis-l-espace_139404 (2019).

54 'Migratory flows in the Ocean: Why do marine animals migrate?' *Fondation Tara Océan* https://fondationta-raocean.org/en/scientific-news/migratory-marine-animals/.

55 Li, J. et al. Zooplankton Fecal Pellet Characteristics and Contribution to the Deep-Sea Carbon Export in the Southern South China Sea. (2022) doi: 10.1029/2022JC019412.

56 Ouertani, N. Le krill: un crustacé aux multiples vertus

thérapeutiques (2017).

57 Krill antarctique | Muséum national d'Histoire naturelle. https://www.mnhn.fr/fr/krill-antarctique.

58 Greenpeace alerte sur le boom de la pêche au krill en Antarctique, *Le Monde*, 13 March 2018: https://www.lemonde.fr/planete/article/2018/03/13/le-krill-un-petit-crustace-tres-convoite_5270007_3244.html.

59 Duchêne, C. et al. Diatom phytochromes integrate the underwater light spectrum to sense depth. *Nature* 637, 691–697 (2025).

60 Cvetkovska, M. Algae use the underwater light spectrum to sense depth. *Nature* 637, 553–554 (2025).

61 Deutsch, J. 7. L'œil de la pieuvre est-il si proche de celui des vertébrés ? *Sci. Ouverte* 67–76 (2017).

62 Le phytoplancton • Les lichens. https://www.plancton-du-monde.org/module-formation/encart_lichen.html.

63 Armstrong, R. A. Adaptation of Lichens to Extreme Conditions. In ResearchGate (2017). doi: 10.1007/978-981-10-6744-01.

64 de la Torre, R. et al. Survival of lichens and bacteria exposed to outer space conditions – Results of the *Lithopanspermia* experiments. *Icarus* 208, 735–748 (2010).

65 Zhong, B., Sun, L. & Penny, D. The Origin of Land Plants: A Phylogenomic Perspective. *Evol. Bioinforma. Online* 11, 137–141 (2015).

66 Chastand, J.-B. Demain, du thon rouge d'élevage ? *Le Monde* (2010).

67 La *Cliona limacina*, prédateur insoupçonné du Plancton – Observatoire du Plancton. https://www.observatoire-plancton.fr/

la-cliona-limacina-predateur-insoupconne-du-plancton/ (2025).

68 Adopt a float. Siphonophores - Les plus longs animaux du monde. *mon océan et moi* http://monoceanetmoi.com/web/index.php/fr/2014-03-27-11-05-40/2014-03-27-11-06-39/2014-03-27-08-26-58/support-des-classes-adopt-a-float/57-fr/ressources/videos/99-siphonophores-les-plus-longs-animaux-du-monde.

69 Ribeiro, A. & Basse, V. Un pour tous, tous pour un: les siphonophores Bionum. https://bionum.u-paris.fr/un-pour-tous-tous-pour-un-les-siphonophores/ (2017).

70 Valo, M. Les cuboméduses tuent une cinquantaine de personnes par an. *Le Monde* (2014).

71 https://www.insb.cnrs.fr/fr/cnrsinfo/des-cotes-la-haute-mer-comment-les-meduses-ont-conquis-locean.

72 Langley, L., How Jellyfish Rule the Seas Without a Brain, *National Geographic* (17 August 2018). https://www.nationalgeographic.com/animals/article/jellyfish-brain-sting-prehistoric-animals.

73 Deux noix d'eau blessées peuvent fusionner et ne former qu'un seul individu. *Courrier international* https://www.courrierinternational.com/article/biologie-deux-noix-d-eau-blessees-peuvent-fusionner-et- ne-former-qu-un-seul-individu_223110 (2024).

74 https://www.bbc.co.uk/news/av/technology-32965841.

75 https://planktonplanet.org/wp-content/uploads/2025/02/Plankton-Planet-PlanktonManifesto_MG_DIGITAL-2.pdf.

76 Agnès b.: 20 ans d'engagement en faveur de l'Océan. *Fondation Tara Océan* https://fondationtaraocean.org/

autres/20-ans-engagement-fondation-tara-ocean/2023).

77 https://barc.ug.edu.pl/2024/06/26/
what-makes-researching-the-seas-possible-tool-1-rosette/.

78 https://fondationtaraocean.org/en/scientific-news/
ocean-science-journey-tara-sample/.

79 Espèces marines: l'heure des grandes découvertes
| UNESCO https://www.unesco.org/fr/articles/
especes-marines-lheure-des-grandes-decouvertes.

80 Thomas-Bourgneuf, M. & Mollo, P. *L'enjeu plancton:
l'écologie de l'invisible* (éditions Charles Léopold Mayer,
Paris, 2009).

81 Argo. https://argo.ucsd.edu/.

82 Qualité de l'eau et assainissement en France (annexes).
Sénat https://www.senat.fr/rap/l02-215-2/l02-215-2.html
(2023).

83 NCEI. NOAA CoastWatch Sea-Viewing Wide Field-of-View
Sensor (SeaWiFS) Level 1A Data. https://www.ncei.noaa.
gov/access/metadata/landing-page/bin/iso?id=gov.noaa.
nodc:CoastWatch-OC-SeaWiFS-L1A;view=iso.

84 https://news.un.org/en/story/2017/05/558352-feature-
grandpa-oyster-offers-example-sustainable-ocean-business.

85 Jephcott, T. G. et al. Ecological impacts of parasitic
chytrids, syndiniales and perkinsids on populations of
marine photosynthetic dinoflagellates. *Fungal Ecol.* 19,
47–58 (2016).

86 Kupec, I. Plankton zombies for Halloween! *EMBL* https://
www.embl.org/news/science/plankton-zombies-for-hallow-
een/ (2022).

87 Plastique: peut-on s'en passer ? ADEME Infos https://
infos.ademe.fr/magazine-juillet-aout-2022/faits-et-chiffres/

plastique-peut-on-sen-passer/.

88 Digital Twin Ocean https://digitaltwinocean.mercator-ocean.eu/.

89 https://science.nasa.gov/earth/earth-observatory/john-martin/.

90 Chami, R., Cosimano, T., Fullenkamp, C. & Oztosun, S. *A Strategy to Protect Whales Can Limit Greenhouse Gases and Global Warming.* https://www.imf.org/en/publications/fandd/issues/2019/12/natures-solution-to-climate-change-chami.

91 Whaling and climate change: the double whammy for carbon sequestration - Université de Montpellier. https://www.umontpellier.fr https://www.umontpellier.fr/en/articles/peche-baleiniere-et-changement-climatique-la-double-peine-pour-la-sequestration-de-carbone (2022).

92 Baleine bleue | Muséum national d'Histoire naturelle. https://www.mnhn.fr/fr/rorqual-bleu.

93 *The State of World Fisheries and Aquaculture 2024* (FAO, 2024). https://openknowledge.fao.org/items/8ab20ccf-1e9d-4ae6-836c-ca770d16da01.

94 Link, J. & Marshak, A. *Ecosystem-Based Fisheries Management: Progress, Importance, and Impacts in the United States* (2021). doi: 10.1093/oso/9780192843463.001.0001.

95 https://www.fao.org/fishery/en/eaf-net/about/what-is-eaf.

96 Où sont passés les anchois et les sardines ? *L'Ifremer en Méditerranée* https://mediterranee.ifremer.fr/l-actu-Ifremer-Mediterranee/Ou-sont- passes-les-anchois-et-les-sardines (2016).

97 Queiros, Q. Mechanisms underlying the bottom-up control

of sardine populations in the Gulf of Lions: insights from experiments and modeling (Montpellier, 2019).

98 Debove, L. 7 000 baleines à bosse mortes de faim en seulement 9 ans à cause des canicules marines. *La Relève et La Peste* (2024).

99 Lesnes, C. De l'Alaska à la Californie, 70 baleines se sont échouées depuis janvier. *Le Monde* (2019) https://www. lemonde.fr/planete/article/2019/06/06/baleines-en-peril-sur-la-cote-pacifique_5472371_3244.html.

100 Barbraud, C. Le succès reproducteur des oiseaux de mer révèle l'impact des changements globaux sur les écosystèmes marins | CNRS Écologie & Environnement. https:// www.inee.cnrs.fr/fr/cnrsinfo/le-succes-reproducteur-des-oiseaux-de-mer-revele-limpact-des-changements-globaux-sur-les (2021).

101 AFP. Il y a 10 ans, une vague de chaleur inédite a soudain tué des millions d'oiseaux marins. *Geo.fr* (2024).

102 Environment, U. N. *Status of Coral Reefs of the World 2020 | UNEP - UN Environment Programme*. https://www.unep. org/resources/status-coral-reefs-world-2020 (2021).

103 10 facts about coral: a marine organism essential to ocean biodiversity | Tara Pacific. *Fondation Tara Océan* https:// fondationtaraocean.org/en/scientific-news/coral-reefs-sentinels-climate-change/ (2023).

104 https://www.mcsuk.org/ocean-emergency/marine-protected-areas/why-marine-protected-areas-are-important/.

105 Convention on Biological Diversity https://www.cbd.int/.

106 Western Sahara Resource Watch | The resource curse. https://wsrw.org/en/the-resource-curse.

107 McKie, R. Scientists warn of 'phosphogeddon' as critical fertiliser shortages loom | *The Guardian*. https://www.theguardian.com/environment/2023/mar/12/scientists-warn-of-phosphogeddon-fertiliser-shortages-loom.

108 Radiofrance. La société NXO Engineering traite des eaux usées avec des microalgues. {2024}.

109 Daheron, N. Un remède aux algues vertes: la start-up Zeni mise sur la phytoépuration par les microalgues. *Presse Océan* https://www.ouest-france.fr/economie/entreprises/startup/un-remede-aux-algues-vertes-la-start-up-zeni-mise-sur-la-phytoepuration-par-les-microalgues-36faee84-effd-11ee-98ed-5193e885a6e0 (2024).

110 Comment des microalgues éliminent les métaux lourds dans les eaux polluées | CNRS Ingénierie. https://www.insis.cnrs.fr/fr/cnrsinfo/comment-des-microalgues-eliminent-les-metaux-lourds-dans-les-eaux-polluees (2024).

111 Grelier, A. Bioteos: des microalgues pour dépolluer l'air. *France Culture* https://www.radiofrance.fr/franceculture/bioteos-des-microalgues-pour-depolluer-l-air-8602340 (2023).

112 Aponiente - El Puerto de Santa María-Restaurant- 50Best Discovery. https://www.theworlds50best.com/discovery/Establishments/Spain/El-Puerto-de-Santa-Mar%C3%ADa/Aponiente.html.

113 *Chef's Table*, season 7, episode 3 – Netflix: https://www.netflix.com/gb/title/80007945.

114 Angel León - World Food Day - Food Heroes detail - FAO. *WorldFood- Day* https://www.fao.org/world-food-day/food-heroes/detail/angel-leon/en.

115 https://www.planctonmarino.com/en/products/.

116 Fortin, P. Algama, des microalgues pour remplacer les protéines animales. *Les Échos* https://www.lesechos.fr/weekend/planete/algama-des-microalgues-pour-remplacer-les-proteines-animales-1915351 (2021).

117 Ciferri, O. Spirulina, the edible microorganism. *Microbiol. Rev.* 47, 551–578 (1983).

118 Swendseid, M. E. *Essential Amino Acid Requirements: A Review.* https://www.fao.org/4/m2772e/m2772e00.htm (1981).

119 Calella, P. *et al.* Antioxidant, immunomodulatory, and anti-inflammatory effects of Spirulina in disease conditions: a systematic review. *Int. J. Food Sci. Nutr.* 73, 1047–1056 (2022).

120 Khan, Z., Bhadouria, P. & Bisen, P. S. Nutritional and therapeutic potential of Spirulina. *Curr. Pharm. Biotechnol.* 6, 373–379 (2005).

121 Hatami, E. et al. The effect of spirulina on type 2 diabetes: a systematic review and meta-analysis. *J. Diabetes Metab. Disord.* 20, 883–892 (2021).

122 Huang, H., Liao, D., Pu, R. & Cui, Y. Quantifying the effects of spirulina supplementation on plasma lipid and glucose concentrations, body weight, and blood pressure. *Diabetes Metab. Syndr. Obes. Targets Ther.* 11, 729–742 (2018).

123 P, M., G, R., M, M., H, P.-S. & A, S. Effect of Spirulina Supplementation on Systolic and Diastolic Blood Pressure: Systematic Review and Meta-Analysis of Randomized Controlled Trials. *Nutrients* 13, (2021).

124 Phansuea, P., Chotchindakun, K., Sahasakul, Y., Phattaramarut, K. & Kuntanawat, P. Effectiveness of an

Arthrospira platensis (Spirulina) Softgel Supplementation on Sleep Quality, Mental Health Status, and Body Mass Index in Mild to Moderately Severe Depression Adults: A Double-Blinded, Randomized, Placebo-Controlled Trial. *Food Sci. Nutr.* 13, e70082 (2025).

125 Moradi-Kor, N. et al. Protective Effects of Spirulina platensis, Voluntary Exercise and Environmental Interventions Against Adolescent Stress-Induced Anxiety and Depressive-Like Symptoms, Oxidative Stress and Alterations of BDNF and 5HT-3 Receptors of the Prefrontal Cortex in Female Rats. *Neuropsychiatr. Dis. Treat.* 16, 1777–1794 (2020).

126 https://www.greentech.fr/wp-content/uploads/2024/04/Greentech_World_EN.pdf.

127 Brevel, Ltd. Launches Commercial Plant for Microalgae Protein Supply. https://www.prnewswire.com/news-releases/brevel-launches-commercial-plant-for-microalgae-protein-supply-302162956.html.

128 https://mialgae.com/.

129 https://www.loebclassics.com/view/pliny_elder-natural_history/1938/pb_LCL418.435.xml?mainRsKey=yNDuY9&result=1&rskey=YD4uUu&readMode=recto.

130 https://archive.org/details/openseaitsnaturaoohard_o.

131 Uluu. https://www.uluu.com.au/.

132 https://glowee.com/aboutus/.

133 FAOSTAT. Pesticides Use in the World (1990 - 2022) https://www.fao.org/faostat/en/#data/RP/visualize.

134 https://nutraceuticalbusinessreview.com/partnership-develops-natural-biocontrol-products-from-microalgae-211195.

135 https://see.systemsbiology.net/wp-content/uploads/files/L5ATeacherResource-primarylit2_PowerofPlankton.pdf.

136 Amaro, H., Guedes, A. & Malcata, F. Antimicrobial activities of microalgae: An invited review. *ResearchGate* (2011).

137 The Continuous Plankton Recorder https://www.cprsurvey.org/home/.

138 Amaro, H., Guedes, A. & Malcata, F. Antimicrobial activities of microalgae: An invited review. In *ResearchGate* (2011).

139 Facing Gaia: Latour (and Battistoni on Latour) https://o-matic.com/migrated/facing-gaia-latour/

Vincent Doumeizel is Senior Adviser on the oceans to the United Nations Global Compact and has also served as director of the Food Programme at the Lloyd's Register Foundation. He has been involved in launching initiatives such as the Global Seaweed Coalition and the UN Global Seaweed Initiative under UN auspices. In 2024 Vincent gathered the forty best plankton specialists worlwide in order to write the UN Plankton Manifesto released at UN General Assembly. A self-described optimist and global citizen, Vincent has in recent years devoted himself to promoting a food revolution and environmental solutions based on sea resources.

He is the author of *The Seaweed Revolution*, as well as a children's book and graphic novel on the subject of seaweed.

Charlotte Coombe is an award-winning British translator working from French, Spanish and Catalan into English, with two decades of experience across a variety of genres including literary fiction and nonfiction. She has translated more than a dozen books by authors including Vincent Doumeizel, Anna Pazos, Frédéric Laffont, Rosa Ribas, Marvel Moreno, Margarita García Robayo, Ricardo Romero, Eduardo Berti, and Abnousse Shalmani. Her translations of short stories and poetry have appeared in anthologies and literary journals such as *The White Review*, *The Southern Review*, *Modern Poetry in Translation*, *World Literature Today* and *Words Without Borders*. She is also the co-founder of the Translators Aloud, a YouTube project that shines a light on translators reading from their work.